NEW DIRECTIONS 30

In memoriam
ROBERT M. MacGREGOR
1911–1974

[**New Directions in Prose and Poetry**] 30

Edited by J. Laughlin
with Peter Glassgold and Frederick R. Martin

 **A New Directions Book**

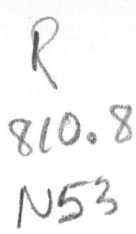

Copyright © 1975 by New Directions Publishing Corporation
Library of Congress Catalog Card Number: 37–1751 (Serial)

All rights reserved. Except for brief passages quoted in a newspaper, magazine, radio, or television review, no part of this book may be reproduced in any form or by any means, electronic or mechanical, including photocopying and recording, or by any information storage and retrieval system, without permission in writing from the Publisher.

ACKNOWLEDGMENTS

Grateful acknowledgment is made to the editors and publishers of books and magazines where some of the selections in this volume first appeared: for Walter Abish, *Statements: New Fiction* (Copyright © 1975 by Fiction Collective/George Braziller); for J. Laughlin, *Poetry Australia;* for Robert Morgan, *Lillabulero* (Copyright © 1974 by Lillabulero Press); for Charles Olson, *Among the Ruins,* edited by Catherine Seelye (Grossman Publishers, Copyright © 1975 by The University of Connecticut Library); for James Purdy, *Esquire* (Copyright © 1974 by James Purdy); for Stanislaw Ignacy Witkiewicz, *The Polish Review* (Copyright © 1973 by Daniel Gerould and Jadwiga Kosicka); for Harriet Zinnes, *Matrix.*

The selected poems included in André Lefevere's "Irritational Verse" were published in the original German by Rotbuch Verlag GmbH, Berlin (Friedrich Christian Delius, Yaak Karsunke, and Peter Schneider), and Klaus Wagenbach Verlag, Berlin (Kurt Bartsch and Johannes Schenk).

Manufactured in the United States of America
First published clothbound (ISBN: 0–8112–0572–x) and as New Directions Paperbook 395 (ISBN: 0-8112-0573-8) in 1975
Published simultaneously in Canada by McClelland & Stewart, Ltd.

New Directions Books are published for James Laughlin
by New Directions Publishing Corporation,
333 Sixth Avenue, New York 10014

CONTENTS

Walter Abish
    Life Uniforms   152

E. M. Beekman
    Ember Days   121

Marvin Cohen
    For a Human Table, See "Lap."
    Laps Commit Lapse.
    See "The Lapse of the Gods."   177

Coleman Dowell
    Victor   34

Mia Garcia-Camarillo
    In Crying's Body   70

Alfred Starr Hamilton
    Seven Poems   174

José Hierro
    Three Poems   112

Steve Katz
    Female Skin   162

J. Laughlin
    It Does Me Good   180

André Lefevere, ed.
    Irritational Verse   75

Robert Morgan
> Climbing 29

Charles Olson
> GrandPa, GoodBye 1

James Purdy
> Summer Tidings 103

Gilbert Sorrentino
> *Art Futures* Interview of the Month: Barnett Tete 12

Stanislaw Ignacy Witkiewicz
> The New Deliverance 133

Tennessee Williams
> Wolf's Hour 161

Harriet Zinnes
> Marie, Marie, Hold on Tight 68

Notes on Contributors 181

GRANDPA, GOODBYE

An excerpt from *Among the Ruins:*
Charles Olson with Ezra Pound at St. Elizabeths

CHARLES OLSON

Edited by Catherine Seelye

Charles Olson was a frequent early visitor of Ezra Pound's at St. Elizabeths Hospital, where Pound was confined for twelve years. "GrandPa, GoodBye" is the final piece written by Olson about these visits, which ended sometime in late spring 1948, two and one half years after their beginning. By this time, Olson had arrived at a reasonably detached point of view, one in which he could see and appreciate Pound the poet apart from Pound the political and economic theorist. This detachment, however, as Olson's record shows, was not easily achieved. Nor was it Olson's final word on Pound: two unsympathetic poems came early in 1950, but Olson thereafter, throughout the remainder of his life, affirmed his indebtedness to and great admiration of Pound. Olson's "Pound file" is to be published by Grossman Publishers, spring 1975, as Among the Ruins: Charles Olson with Ezra Pound at St. Elizabeths.

C.L.

Copyright © 1975 by The University of Connecticut Library

1

Time is in his conversation more often than anything else. I said to Lowell[1] the other night: "There is a haste in Pound, but it does not seem to be rushing to any future or away from any past." It is mere impatience, the nerves turning like a wild speed-machine (it is how he got his work done) and, more important, an intolerance of the mind's speed (fast as his goes), an intolerance even of himself. For he is not as vain as he acts. "30 yrs, 30 yrs behind the time"—you hear it from him, over and over. It is his measure (and his rod) for all work, and men. His mind bursts from the lags he sees around him.

He speaks as though he found himself like retarded when he began. Apropos Ford (F.M.), he said to me once: "From the intellectual centre, 30 yrs start of me." He elaborated on it, another time: "Ford knew, when I was still sucking at Swinburne." (He credits Ford somewhere as the one who formulated the proposition, c. 1908: verse should be at least as well written as prose.[2]

Now he puts all little magazines aside, or questions about someone's work, or presentation copies, with a jerk: "I can't be bothered. I've worked 40 yrs, I've done mine, and I can't be bothered trying to find out what the new candidates don't know."

"When I was a freshman . . . ," he started once.

2

If I were asked to say what I thought was the pure point of the Old Man as poet I would say the back-trail. The more familiar observation, that it is his translations on which his fame will rest, is a step off the truth. They are a part of the career he has built on remembering, but the root is a given of his own nature. The lines and passages which stand out, from the start, capture a mood of loss, and bear a beauty of loss. It is as though Pound never had illusion, was born without an ear of his own, was, instead, an extraordinary ear of an era, and did the listening for a whole time, the sharpest sort of listening, from Dante down. (I think of Bill Williams's remark: "It's the best damned ear ever born to listen to this language!")[3] He said to me one of the first days I visited him, when he was in the penitentiary part of the hospital, what he has now come to call the Hell-Hole to distinguish it from his first detention cell at Pisa, the Gorilla Cage (where he felt he had been broken), "Among the ruins, among the ruins, the finest memory in the Orient." (He sd, "Orient.")

His conversation, as so much of the Cantos, is recall, stories of Picabia, Yeats (Willie), Fordie, Frobenius, Hauptmann,[4] of intelligent men, and it is as good as you can get. I was never made so aware of what a value he puts on anecdote as recently, when I returned from seven months on the Pacific Coast, and he jibed me, to what use I had put my time. I got around to my adventure in Hollywood, and young Huston's story of Jack Warner and the Whale. It pleased Pound much, and, as he sometimes does, wagging like an old saw, he says to Dorothy,[5] "Well, seven months, one story, not bad, not bad at all. There aren't so many good stories after all." He's a collector—what's that line he had in the Cantos and took out, about scrap-bag, the Cantos as same?[6]

I dare say these are now commonplaces of Pound critique, but I don't think it has been sufficiently observed, if it has been observed, how much his work is a structure of mnemonics raised on a reed, nostalgia.

3

Edward Dahlberg has it, in *Do These Bones Live,* that ennui is the malaise of the life of our fathers, modern life I think he says, and in reference to Dostoevsky, and his women.[7] Is it far-fetched, or too easy of me to take Pound's haste, and this vertu of his verse, as born of same?

It is hard for me to do otherwise, having known him. His power is a funny thing. There is no question he's got the jump—his wit, the speed of his language, the grab of it, the intimidation of his skillfully-wrought career. But he has little power to compel, that is, by his person. He strikes you as brittle—and terribly American, insecure. I miss weight, and an abundance. He does not seem— and this is a crazy thing to say in the face of his beautiful verse, to appear ungrateful for it—but I say it, he does not seem to have inhabited his own experience. It is almost as though he converted too fast. The impression persists, that the only life he had lived is, in fact, the literary, and, admitting its necessity to our fathers, especially to him who had such a job of clearing to do, I take it a fault. For the verbal brilliance, delightful as it is, leaves the roots dry. One has a strong feeling, coming away from him, of a lack of the amorous, down there somewhere. (I remember that Robert Duncan, when he returned to California from his cross-country pilgrimage to Pound, spoke of this, was struck by it.) E.P. is a tennis ball.

(When I think of what I have just said, and of those early poems!)

An attempt like this is such a presumption: said Frederick of Prussia, "Every man must save himself in his own way." You might say I am offering these notes out of curiosity, on the chance that they may or may not illumine him. He can stand it. He's no easy man. He has many devices. And he's large. I'm not sure that, precisely because of the use he has put nostalgia to, and the way he has used himself, he has not made of himself the ultimate image of the end of the West. Which is something.

(I am reminded of a remark of—I don't remember which, so help me, Lao Tse or Confucius!, that a shrewd man knows others, a true man illuminates himself.)[8]

Wait. I think I've got it. Yes, Ezra *is* a tennis ball, does bounce on, off, along, over everything. But that's the outside of him. Inside it's the same, but different, he bounces, but like light bounces. Inside he is like light is, the way light behaves. In this sense he is light, light is the way of E.P.'s knowing, light is the *numen* of him, light is his way.

And that is why he goes as he does, and why he is able to make his most beautiful poem of love later than "Cino,"[9] not so different from "Cino" either, but vastly more complete, and make the whole Canto—I am referring to XXXVI—a straight translation from Cavalcanti, the Cavalcanti itself one whole extrapolation of love in terms of light, and drawn, in its turn, straight from Grosseteste's essay on the "physics" of light.[10]

Thus also Pound's fine penetration of Dante? (I am thinking of those images of light in—and images of Beatrice, too—in the "Paradiso," is it, or is it at the end of the "Purgatorio"? Anyway, those which Eliot has made such a poor use of.)[11]

Maybe now I can get at this business of *amor* as of Ezra, and get at it right. It isn't a lack of the amorous, perhaps, so much as it is a completely different sense of the amorous to that which post-Christian man contains, to that which—to be most innocent about it, and properly relative—the likes of Duncan, say, or myself may feel.

(Or the likes of Bill W.? I am struck by the image of "fire" in "Paterson." Maybe fire is the opposite principle to light, and comes to the use of those who do not go the way of light. Fire has to consume to give off its light. But light gets its knowledge—and has its intelligence and its being—by going over things without the necessity of eating the substance of things in the process of purchasing its truth. Maybe this is the difference, the different base of not just these two poets, Bill and E.P., but something more, two contrary conceptions of love. Anyway, in the present context, it serves to characterize two differing personal *via:* one achieves its clarities by way of *claritas,* the other goes about its business blind, achieves its clarities by way of what you might call *confusio.* At which point I quote Chaucer's Cock to his Dame in the middle of the night on their perch. And quit the whole subject.

> For al so siker as *In principio,*
> *Mulier est hominis confusio,—*
> Madame, the sentence of this Latyn is,
> 'Womman is mannes joye and al his blis.'[12]

4

A long time ago (what, 25 years?) Pound took the role of Confucius, put on that mask, for good. I'm sure he would rest his claim not, as I have put it, on the past, but forward, as teacher of history to come, Culture-Bearer in the desert and shame of now. (I don't think it is possible to exaggerate the distance he goes with this notion of himself—at the end Gate of the last Canto Confucius is to be one of the two huge figures standing there, looking on.)

It is all tied up with what he calls a truism, London, 1913: are you or are you not, a serious character?[13] Then it was Gaudier, Lewis and himself. Now it is the Major (Douglas), Lewis and himself.[14] He sd that the other day, when we were talking about Bill Williams. Bill had just been taken to the hospital, and one of the new candidates on 2nd Avenue had had a letter from Bill repeating over and over again, "I have nothing to say, nothing to say." E.P. didn't pay this much mind. Sd he: "Bill has always been confused. He's one of the reasons I make so much of race. It's hard enough for a man to get things clear when he's of one race, but to be Bill! —french, spanish, anglo, some jew from Saragossa. . . ." And though I left Pound that day and shall not see him again, he went on to say something which is true, that what has made Bill important is that Bill has never sd one god damned thing that hasn't first circulated entirely through his head before it comes out his mouth. (Bill never faked, and that's why he has been of such use to all us young men who grew up after him. There he was in Rutherford to be gone to, and seen, a clean animal, the only one we had on the ground, right here in the States.)

5

Pound makes a lot of the head. You can tell that he thinks of it as a pod, and most of the time comes to the conclusion that very little rattles around inside most known pods. The Major, Lewis and himself. The Major, Lewis and E.P. "Lewis," says he once, "Lewis most always gets things wrong. But he gets somethin'!"

At that point his wife got the conversation off on Eliot, and Hulme,[15] confusing 1914 with 1913. Pound did not say that Eliot was not a serious character, what I gathered was that Pound had hoped the Possum might turn out to be one, in 1914. He then went on: "The war got Hulme too young. He used to spend three hours a week with the Bloomsbury gang, Ashley Dukes. . ." he reeled off others "and we'd laugh at him, Gaudier, Lewis, I" and you could see Pound giving it to him. "But when he was gone, when he wasn't there to do it, we caught on, we saw what he'd been up to—he'd been beating that gang over the head hard enough every time he saw them to keep 'em in line. After he was gone, they went on the loose, there was no one to tell 'em . . . And look! . . ." Grampaw was giving a lesson.

Another lesson that day was his story of (Hunecker? no, some H, some citizen whose name I missed—they come flyin'), who was listening to Sturge Moore[16] do a lecture, or a reading, in one of those rooms which are built like an operating room, the seats steep and down like a bowl, and who was wishing that Moore would get on with it, and done, and holding his head in his hand, when a character in the row above him taps him on the shoulder, and sez, "Friend, don't feel so bad, what makes you think we're here to enjoy ourselves?"

He is driven, E.P., to get on with things, his things, the "serious." He sd to me another time: "I can't git it into yr american heads that the earth makes one complete circuit once EVERY 24 hrs."

Another crack: "I thought you might be a serious character when I read that labor-saving device of yrs on H. Melville. But that was 2 yrs ago, bro."

6

When I happened to ask him what Frobenius looked like, I stumbled on the fact that Pound sees himself as a cat man. Both he and the wife quickened to the question. Glances were exchanged. He took it up, first: Sd he, "There were three men"—it was almost as though he said, once upon a time—there were three men who might have come from the same genes. Frobenius . . ." DP interrupted, and started to go on to illustrate the differences between EP's face and Frobenius'. Then Pound was up on his feet, showing how F stood, back on his heels (P himself seems always on his toes, or did, before he thickened up from lack of exercise his second yr at SLiz, around the belly and below the back) with his arms out and his hands stuck in his belt, "He must 'ave got it from Africa," sez Pound (from the exigencies of the place, I judged, but the point being, all around the three of us, that E.P. never in his life would have spent the years that Frobenius did pushing around Africa). DP went on making with her hands, seeking further to show the differences—F not so wide at the temples . . . a V for F's beard, when all of a sudden old Ez lets it out, sort of half to her, all to me, "cat family, cat family," making with his head, and getting that fix in his oye![17]

And the 3rd of the same genes? "Barkoff, Barkoff" (EP with obvious delight) "now Chief of protocol, Moscow. (Pound had mentioned this Barkoff (Barkov, I suppose) several times two years earlier, particularly at the time when he was riding the idea that the U.S. Govt. could have saved itself a lot of trouble—and the world a lot of war—if it had allowed Old Ez what he asked for, the chance to pick up Georgian—"a week, 10 days"—and get over there and talk it out straight with Joe, in his own tongue. I gathered then that what would have made it all so simple was Barkov's presence, to rig the talk, once the USG wised up. Pound seems to have known Barkov years before, in Italy, perhaps, I don't know.)

7

Madame and Sir Pound also managed to get this across one day, that they had recently learned, "through Agnes,"[18] whoever she was, that Frobenius had said of E.P. before he died, "He is worth three Oxfords with four Cambridges piled on top."[19]

And it felt good to hear it, for these praises are old Ez's OM's[20] and Nobel Awards. Joyce, apparently, was not a man to give praise out easily, and Pound told me once, with what pleasure, the only thing he ever got out of Joyce, this comment one day when he had read something new, "The sleek head of verse, Mr. Pound, emerges in your work."

He has a story of Joyce and Hauptmann, Joyce, when young, had translated one of Hauptmann's plays.[21] After Pound had met Hauptmann, Joyce put a copy of the play into Ezra's hands and asked him to get Hauptmann to autograph it for him. "And," says Ez, "he didn't write off one of those inscriptions"—the intervals, that way Pound gets gesture in between the letters even, showed that he was thinking of the way he does it, dashes them off—"he took it away, came back three days later, gave it to me, it read, to JJ the best reader this play ever had, and I shipped it off." At which point he cries out at the top of his delight, "He sure sat on that one till it hatched!"

8

Another of his comforts, is to see himself in the part of the Old Man, "Grandpa," now that Mary has made him one. Sd he the other day, mistaking, I think, the significance of what I was calling to his attention, the coming to Washington (as tho to Rome) of the "Poets" (Eliot at the National Gallery the spring before, which was like a laying on of hands, the coming into existence of an american poet laureate;[22] the creation, and the turnover, of the "poets of Congress," the Consultants; and, at the time I was speaking to him, the run of a series of readings at the new Institute of Contemporary Arts, Spender, Tate, Lowell & the other "Jrs"). Says the O.M.: "Ya, they'll all be coming to old Ez on his deathbed and telling him he was right"

and, when I was telling him about the way the people in the streetcars were shaking their heads like a bunch of dipping birds over a black new headline out of the Ant-Hill, "If they'd only listened to old Ez in the first place!"

"right,
                    right

from the start[23]

(1948)

NOTES

[1] Robert Lowell, who was in Washington in 1947–48 as Consultant in Poetry at the Library of Congress.

[2] Pound first credited Ford with the proposition in 1914 when he wrote: "I cannot belittle my belief that Mr. Hueffer's realization that poetry should be written at least as well as prose will have as wide a result . . ." in a review ("Mr. Hueffer and the Prose Tradition in Verse," *Poetry*, IV, June 1914) of Ford's *Collected Poems*. Ford's belief that poetry must be divested of its "literary" qualities was expressed by him in an essay published as the Preface to *Collected Poems*.

[3] See Williams's essay, "Excerpts from a Critical Sketch: A Draft of XXX Cantos by Ezra Pound," in *Selected Essays of William Carlos Williams* (1954), in which he refers to Pound's "fine ear"; and *The Autobiography of William Carlos Williams* (1951) p. 225, where he says, "A man with an ear such as his, attuned to the metrical subtleties of the best in verse. . . ." Williams continued to praise Pound's competence even when he was disturbed by the "callousness" of Pound's political views. See his statement that Pound "possessed . . . the most acute ear for metrical sequences, to the point of genius, that we have ever known" in Charles Norman's *The Case of Ezra Pound* (1968), p. 84. It took Olson to say "that the great 'ear / can no longer hear!'" in "I, Mencius, Pupil of the Master . . . ," a sentiment he did not long retain.

[4] Francis Picabia (1879–1953), French painter, early Dadaist, later developing a highly personal and imaginative style. Leo Frobenius (1873–1938), German ethnologist and explorer whose primary interest was Africa. Gerhart Hauptmann (1862–1946), German dramatist, exponent of the school of naturalism.

[5] Dorothy Shakespear Pound, Pound's wife.

[6] Pound's first version of Canto I (revised several times after its appearance in *Poetry*, X, June 1917) is as follows: "Hang it all, there can be but one *Sordello*! / But say I want to, say I take your whole bag / of tricks / Let in your quirks and tweeks, and say the / thing's an artform, / Your *Sordello*, and that the modern world / Needs such a ragbag to stuff all its thought in." Fifty years later: "'I picked out this and

that thing that interested me, and then jumbled them into a bag. But that's not the way,' he said, 'to make'—and here he paused—a *work of art.*'" Noel Stock, *The Life of Ezra Pound* (1970), pp. 457–58.

[7] Dahlberg writes, "The Western man of ennui is the heir of the fat dregs of Hamlet's sexual sickness. 'O God! I could be bounded in a nutshell . . . were it not that I have bad dreams' is the malady of boredom of the European, of Pascal, Dostoevski, Tolstoi, Stendhal," in *Can These Bones Live* (1960), p. 156. In the same essays are found "At the nethermost core of history, and at the underside of war and poverty, lies tedium. It is the grand malaise of the Western world" (p. 21), and "Thoreau . . . erected in *Walden* the Western Fable of Ennui" (p. 127).

[8] "He who knows others is learned, he who knows himself is wise." *The Wisdom of Laotse,* trans. and ed. by Lin Yutang (1948) p. 199.

[9] "Cino" first appeared in *A Lume Spento* in 1908.

[10] See Robert Grosseteste, *On Light* (1942), and Carrol F. Terrell, "A Commentary on Grosseteste . . . ," *Paideuma,* Vol. 2, No. 3 (Winter 1973) pp. 449–70.

[11] They are in "Paradiso." Eliot's images of light to which Olson refers are probably those in "Burnt Norton" or possibly those in "Choruses from 'The Rock.'"

[12] *The Nun's Priest's Tale.* (Text from Robinson, p. 242).

[13] Pound is probably referring to his essay "The Serious Artist," first published in *The New Freewoman* (Oct.-Nov. 1913), republished in *Pavannes and Divisions* (1918).

[14] Henri Gaudier-Brzeska (1891–1915), French sculptor, memorialized by Pound in *Gaudier-Brzeska: A Memoir* in 1916 (reissued, 1970). Wyndham Lewis (1884–1957), English writer and painter, leader of the Vorticist movement with which Pound was associated. Major Clifford Hugh Douglas (1879–1952), a Scot social economist who introduced the theory of social credit which Pound championed.

[15] Thomas Ernest Hulme (1883–1917), English philosopher with a strong interest in aesthetics. Pound included Hulme's "complete poetical works" in *Ripostes,* published in 1912.

[16] Thomas Sturge Moore (1870–1944), English poet and artist.

[17] Hugh Kenner reports that when Frobenius and Pound met, "each was astonished by his physical resemblance to the other." *The Pound Era* (1971), p. 507.

[18] Agnes Bedford, a musician who collaborated with Pound on *Five Troubadour Songs.* She later helped Pound with technical matters when he began to compose operas.

[19] Olson uses this sentiment three years later in "Letter for Melville": "a very great man, said / of another—who never learned a thing / from Melville—worth / 'five Oxfords on ten thousand Cambridges'!"

[20] The Order of Merit is a highly sought British honor conferred for special distinction. Eliot was awarded the Order in January 1948.

[21] The play in question here is *Michael Kramer.* See *Letters of James Joyce,* Vol. 1, p. 398; Vol. 3, p. 415. Joyce also translated Hauptmann's *Vor Sonnenaufgang.*

[22] Olson might have been less derisive had he known of Eliot's financial position: "'I recall that a number of years ago when I had less

money than I have now, if I was in New York and wanted to go down to St. Elizabeths Hospital in Washington to see Ezra Pound, I had to manage somehow to get a date to read in order to foot the bill.'" William Levy, *Affectionately, T. S. Eliot,* (1968), p. 64.

[23] For three years, out of key with his time,
   He strove to resuscitate the dead art
   Of poetry; to maintain "the sublime"
   In the old sense. Wrong from the start—

> Ezra Pound
> "Hugh Selwyn Mauberley"

ART FUTURES INTERVIEW OF THE MONTH: BARNETT TETE

An excerpt from the novel *Synthetic Ink*

GILBERT SORRENTINO

The following interview with Barnett Tete, self-made multimillionaire, art collector, connoisseur, raconteur, and cynosure of the international art world, was the culmination of more than two years of negotiations between him and the editors of Art Futures. *Mr. Tete was understandably concerned that his rather esoteric and idiosyncratic views on art and artists might be misunderstood by the public. However, the interview, conducted at Mr. Tete's Long Island estate, "Les Shekels," turned out to be one of the most provocative, compelling, and lucid that it has ever been our pleasure to print. We offer it to our readers as proof that the dedicated amateur can have an understanding and love for his avocation as profound as that of the most committed professional expert.*

QUESTION: First of all, Mr. Tete, exactly how did you make the money that has enabled you to become America's most famous collector and all-around patron of the arts?

ANSWER: There is no point in this, after all. Let's just say that I am known as King Corrugated, and let it go, right? Even a scarecrow has dignity.

Q: Check. You were, according to your biographers, a poor storekeeper and you got this mysterious feeling from art, if I'm remembering right?

A: That's about it. That feeling that the all that is humanly sacred was there. My wife helped me a lot in this feeling, I'll say that for her.

Q: Few of us agree with all you've said about the joys of collecting. Would you care to comment on that?

A: I really don't—I mean that you, I don't mean *you* you, I mean in general, those who have pettiness in the heart, anyway, they can shove it.

Q: I am curious about your early years of poverty and the like . . .

A: The usual squalor. The Cantor-Jessel syndrome. You'll note vague traceries of my banjo eyes and speech impediment. Of this I beg you. Look, a dollar made is in the hand, you can't knock it. Opportunities abounded.

Q: Your parents encouraged you?

A: Leave them out of it, Wisenheimer. Does a fledgling throw a rock from its nest? Look, I'm a busy man. I deal with life as it comes to me aquiver. Few agree and fewer still get the thrill of understanding. Art is my game, O.K.?

Q: Would it be fair to say, as a recent poem does, that you're a lousy guy?

A: There's a streak of meanness in that so-called poet. He thinks he's an artist but there can be little doubt and less thought that he is not at all interested in the beauty that has not yet come into the world.

Q: You're trying to indicate that that is your kind of beauty I take it?

A: One must be the master of warm feelings anent arts in general.

Q: Yet with all your great understanding and compassion for artists and the like you are still the king of the box world. How do you explain this?

A: Tintoretto or somebody said that all the world is a box. The word has a good shape. As King Tete, an appellation of which I justly and proudly embrace, my peers and inferiors hold a high opinion, young man. I have warehouses full of objects de art! I did not get all these riches by fooling around. I am at once utterly raw and unbearably gross.

Q: I had no idea that Tintoretto said that! Do you feel that the Cubist Movement must be viewed then in a new light?

A: No.

Q: What do you feel about Cubism, by the way?

A: I don't understand it. Life is a juice. You cannot, to your immortal peril, lock yourself up in a room while the world, in all its terror and gorgeous looks, whines by outside. Those who crawl toward the light will break the shell. Others get pleurisy and TV back.

Q: Is there a reason why you have your renowned fear of water?

A: For the nonce in the past I fell or was pushed into a vat of molten cardboard. It seems like yesterday. My wife shares my interests in this fate with a sickening sort of regularity.

Q: Your wife . . . does she help you in your purchases?

A: Rarely. She likes to potter about the house and also make the most colorful and gay piñatas. You've seen them at Christmas parties. Life is quirky.

Q: That is an odd view for an American. Would you expatiate?

A: I feel that opposites are one and that Melville never made himself naked before all men. The shell breaks and the crab always rots.

Q: A kind of enlightened Platonism?

A: Insofar as it deals with the explosion of your sensibilities.

Q: What about Kafka?

A: A little too Jewish for my taste. I want an oval world. I used to understand that but a world of surfaces seems to be the "modern" ticket. You recall the hula hoop? In its mastery I've always felt the pastiche of the cold breath of Prague. Often I count my money in glorious crassness.

Q: You feel that he is gloomy then?

A: Of a quintessence. Unlike our friends, the Chinese, who have the intellectual honesty to kick old men and schlepp those big banners, he turned inward. I admit however that he has reified certain concepts, but who cares about narrative?

Q: What about your *own* enormous influence though?

A: What is influence but the shadow of a cloud? How shall I speak of a yellow flower trembling slightly in the breeze? You could get sick from it. You wait around, shifty-eyed, then pounce on the lucky bastards!

Q: I'm thinking of the sudden demand for Krabo's work after you bought his early crayola things . . .

A: Oh, I see. Well, if I may quote the ancient Hebrew proverb, "If I am only for me, who will be?" That old saw has been passed down from father to son for generations like a dark cloud blots the sun, but who's the worse for it? Krabo is all right. My wife discovered him, by the way.

Q: I think I see Mrs. Tete over there in the rose garden or something.

A: She is all heart. That is her little boots protruding from a clump. Krabo did a beautiful drawing of her, "Lady Tete in the Bushes."

Q: Isn't that the famous one without a picture plane?

A: Well, famous, I don't know . . . nobody really noticed. A petty mind is a hobgoblin. Actually, Bart Kahane knew.

Q: Wasn't Kahane blind at the time that picture was exhibited?

A: Isn't that remarkable? Well, Krabo might as well be blind. He thought he was too smart to be a sculptor, that's what books will do.

Q: Is Krabo still working in crayola? I've heard talk that he's been doing some experimental things with rust.

A: Fools often come into the light of day when it seems darkest. It's no secret that Krabo and I are finished. It's a matter of what they call the experience, no? Look, the Rocky Mountains still present their great lesson. He made his pile. Let him lay in it.

Q: I'm interested in what you've said about Mrs. Tete and Krabo. Does your wife often urge you to buy an unknown—does she have that "eye"?

A: Often it seems to be a contingency. What else can one declare? When she reaches into the hive of her angers and enthuses, she plucks out the big ones. Three lemons! Three plums! Personally, I have been a crab all night most nights. But she is loaded with transparencies. One thinks of Thaddeus Stevens wisely wandering by old synagogues.

Q: I am quite taken with your literary allusions. Are you interested in literature?

A: I'm mad about good books. A favorite, of course, is Lawrence. I consider him to be the great creative genius of our age. You know, of course, that he could swim and that he was not averse to sweet marmalade despite his protestations or affectations. He stepped lively and pointed out the uninterrupted significance of green things flying diagonally. Who can deny or dare forget that he freed the country from rocks?

Q: There are some who think that he was a morphodite. Comment?

A: That is mere legerdemain.

Q: What are you currently reading?

A: At present I am ensconced with *Thou, Thee!*, Marion Gusano's new book of verse. Or dare I call it verse? The word seems so squeamishly small and clerkish. In Poetry's great face somewhat, how shall I say, embarrassed? The language is full of dirt as well as being full of shit. Self-challenge repletes throughout. Plus trembling risk also. A kind of Blakean *mot* in its lack of attention to this and that, if you are following my drift. A great heart, in all events, rings in every line. Also, I'm reading *The History of Acrylics* by Benabou. This latter is, I fear, simply a matter of a busman driving to Newcastle.

Q: Generally speaking, do you feel that Poetry has affinities with the arts of painting and sculpture?

A: I am as sure of this as I am of the sores of life and that the sun reveals the color of the garbage. I am not one of those who will not let the ocean lave him. Tantric sounds awaken the goddess that lies coiled at the base of my body. I adore the meeting, the ongoing dialogue that spells wonder! It is no accident that Max June, one of my more servile protégés, showed collages last spring that closely follow the newer developments in certain verse forms, of which I don't want to explain about them just at present, so thrilling are they. Suffice it that they have the power of the Atlantic bunking into South Carolina.

Q: Someone seems to be in pursuit of your wife.

A: Yes. That is Shreve, the gamekeeper. Pastoral soul! His great heart rages, but he has a certain difficulty in making himself understood. You'll note how like children they are trampling the hollyhock bushes, which seem to be in danger of imminent liquefaction. Yet the garden fairly sings with their scraping.

Q: You would say then that Shreve is the fabled "natural man" come true?

A: The language I long for is closed to me, yet I would have to say yes. When he first arrived, a shambling lout from the country, he used to eat his belt. Now Mrs. Shreve, who knows the value of the last color golden on the white birches, has got him to the point at which he asserts the light. Yet he is still ferociously visionary, and ignores personal hygiene.

Q: Your interest in the primitive has always been a source of fascination. I'm thinking of Louis Henry . . .

A: Oh, yes. I'm glad you brought Lou Henry up. In those watercolors he, I am certain, has found himself whole and wholly too. Somehow, he could not pass beyond sheer fabric in his verse. You'll recall the *Lobster Lays*, of course. Symptoms of bitterness, which is literalness, pervade the work at all points, and what, I ask you, is at the root of that? But don't ask, I'll tell you. It is a resolute hollow old barrel, nothing more or less.

Q: But you feel that in his painting . . . ?

A: I feel that he has found himself, certainly. There are many who laugh, who say that this work of Lou's is mere satiric appositeness. Others say that there is no air in it, that he falls into his reflection. Is life an acid? I think we must agree to destroy the pigeon holes. Everlasting psychic charge is my sense of it all, take the moon made of paper. I don't think it's an accident that Henry's work is a continent big enough to take care of sourness and snottiness. I've often seen him feeding Chinese to his inamorata while his wife looked on agape.

Q: Didn't his wife do the cover for *Lobster Lays*?

A: Sheila? Yes. Of course she is an artist in her own right, something like a poem with a novel trying to get out. Few think or e'er forget that when Rimbaud understood this he gave up and left town. It was a sweet journey. So goes beauty that retires from the light.

Q: What do you see as the so-called *lumpen* qualities in Henry's aesthetic?

A: That's Jungian terminology and I don't agree. Are you maybe thinking of gloom and misery everywhere? But a moment's reflection will show that Henry's murky pictures are of a despair. Look, his people are absolutely dead. And if all true art is vital, then what? I mean, the impending transformation of all shapes. Is this a dichotomy, or what?

Q: You seem to be hinting at a kind of mysticism?

A: I don't know from this. A man has commitments, I call your attention to great predecessors of whom I continually think. There are those who drain reality of its imaginative power. Exhilarating as this may be, it falls far short of the ant on the bathroom floor. The world may be a wisp of shadow yet men place their papers in the safe-deposit box. Speaking of Jung, of this latter statement even he might have cracked a smile. He was not given to levity as are none of our great souls yet rumors persist that he was guilty of the Gogolian flaw. Take it with a grain of salt or not, the good Herr Doktor was right about San Francisco. And if *he* was a mystic I'll wolf my homburg.

Q: Has collecting artworks and befriending artists been enough for you?

A: What is "enough"? Rubens once said that he holds who cries enough, but you cannot trust the Germans. There have been, of course, lengthy bawls in the night, but they have been more than compensated for by compelling results. One thinks of the moment of art when there is no more to do than plunge the hands in the pockets. Art is an opening. Well, perhaps an act of faith that there is something on the other side. Whatever it is, it calls for a suspension of the brains. Is an acid a juice? I think not.

Q: That's very revealing in light of your discovery and championing of Moss Kuth's "invisibles." Would you like to comment?

A: When I first saw Kuth's incredible work, I didn't know whether he wavered or walked alabaster to his destination. Such was my amaze! I thought, "Can I talk to you who I must create?" It so turned out that though his visage was grim a sweeter man could not be long imagined. For an artist, he's got a great sense of manners. They will come no more. Kuth, a member of that heavenly company if there ever was one, has passed beyond the insular concern with making and, more importantly, beyond the vaguely intellectual and "literary" concerns—epistemological, I insist, if the word has any meaning at all—of what some of the fatheads term "destruction." Call it visual silence. There is not only nothing there, nothing ever was there. If this isn't the *crème de la crème* of the avant-garde, I ask you. If it is true that the artist can never close his anus, it is equally true that thousands of elves keep watch thereby. Yet how many laugh! In Kuth there is the fourth dimension made flesh. Gibe if you will, but fattest peacocks fall before the sun. Mumblings suggest that this is a red-hot property, but I know what I like. I have cast my bread upon the waters and it has come back apple pie.

Q: Kuth's work at first evinced many harsh and uncomplimentary critiques. What changed all that, would you say?

A: Well, early Kuths curled at the far edges, as you can well imagine. Is there an adventurous soul near or far who will stand still for that? You of course realize that I use the word "adventurous" advisedly, ha ha. Equally accumulated senses of composition were largely displaced. Only a "sap" would pull out his bulging wallet. In sum, yours truly. When shekels flow are dealers far behind? But seriously, folk, Kuth is the wave of the future. What can be more hotsy-totsy than to allow the planar to yield to the linear? After great planes, a formal boredom comes. Then Kuth, shedding those crocodile tears for which the West Germans have long applauded him, eschewed even the linear. Other painters, despite inner feelings of disgust, mouthed the famous phrase, "We will not spit in the soup." When even the paint and canvas disappeared serenity reared its head, like Venus from the river. Call it what you like, Kuth has a sense of manners, you may well impute, despite all.

Q: Yet Kuth seems to imply in a recent interview that he misses tangible objects.

A: Let me hasten to say aught. That is an Aristotelian fancy. Since then the artist has grasped the meaning of the orgone and found it good. When a painter breaks ground, you will pardon the expression, he finds conditions of contemporary environment, common vocabulary, and conceptual "aesthetics." Perhaps art, like life, is also a juice, though one shudders to think so. I've suggested that Kuth should cut out all this reading business and get back to basics. Books spoil the sense of visual excitement that the painter has buried, like a great golden gift, in his peepers. And no matter what, when all the critics have been heard from, all the backs bitten, the art histories written, even the company especially pleasant? Personally, I refuse to be swayed by such gewgaws.

Q: In sum, then, would you say that "invisibles" have the effect of an expanding pressure against a limit of edge?

A: An excellent mélange of verbiage, yes.

Q: Some women artists have said that your taste runs to those painters whose work is most redolent of male supremacy and so on. Is there any truth to that?

A: Look, I like women. God could not be everywhere, that's why He made mothers. Their boxes, as well as ours, crackle with Reichian blueness. When they stumble, do they not trip? Think on Miz Tete, one of the county's finest examples. I would as soon underpay a woman as a man, that's where my heart is, though my seamed face may belie it. There are few prostitutes left with no sense of the people's needs. This I applaud with selected encomia. And though there are many truly rotten male artists they hold no edge, numerically speaking, over the distaffs. In my position I do not like to get into these beefs. Let's drop this frou-frou and assault, rather, the great questions that e'er bring us up short with their deepness, oke?

Q: I'm glad you feel that way. As a solid appreciator of art, what do you feel about surface—I mean, as opposed to what lies beneath?

A: That is a tough question and I can only pretend a complete answer according to my lights. I'll give you various contours, how does that sit? I can say that the distant call of birds is like blue coins of disaster, but will that prevent even one asymmetry? The thought perishes. You'll recall the work of Benny Dredger. If that was not the very posy of a poetic aesthetic, I'll savage my derby! You may think that I am here equating painting with artifact and I do confess that his work, the sheer glittery and tinsely "trappings" and so on did indeed at one time have me jumping up and down. Though many have called him a miniaturist, others have not. It is that silence that affronts. Equally, my purposes do not understand anything more than what they propose. Certainly that seems fair. To juxtapose the interior with the exterior—it was the way of all the great Geminis. Think of the boiling face of Himmler and the dead pan of Al Jolson. Who would think that both had souls as succulent as chicken shaked and baked? True it is, but then I have always loved Whittier.

Q: You are saying that surface is artifact, and the interior . . . ?

A: In the purely craft sense, to make something is a later idea. I like the poor but they can't buy things. Your questions offend and I feel the beginning ache in my corns. Mack Jackson once said, "Sit quiet in white spaces." At the time the moon was all aglow. Its wont demanded it. Even I, a humble box tycoon, may plunge into the Self and find happiness. Art must be alive in one sense or another! How few believe this credo. Hopeful and pompous, however, I plod on, one of my few delights the knowledge that a wigwag of my little finger can set the rabble on a bore. My wife often quakes at my black moods, then breaks into her swell laugh when she sees that it is only the Self that is coming up, belch-wise. The painter must formalize that impetus or poopies on him.

Q: Speaking of your wife, she and Shreve seem to have disappeared, yet I'm sure I hear their voices . . .

A: There is a fat chance that they are straining after truth in the gazebo. But it's not our purpose to dwell on the life of the spirit, is it?

Q: Of course not, sir. I wanted to ask you about your recent purchases of ink drawings by Jigoku Zoshi. Have you had an interest in Eastern art for a long time or is this a recent enthusiasm?

A: Eastern art? What a remarkable idea! I've never quite thought of it that way. You've come a long way from St. Louis! I must admit I am charmed by the appellation. It's got class. Actually, I am celebrated worldwide as having the mind of your average slob, yet a grand here and a grand there sweetens one's occasional gaffes. Eastern art is of a mystery, you can spell that with a capital M. Who am I to cipher? A man, like all men, yet one who was able to take advantage of a stupid partner in his salad days. At present the honest wretch can be found selling pretzels outside Moskowitz & Lupowitz. Often, in the gloaming, when the breasts of the housewives are heavy with salt, I drop a tear when I fondle a Zoshi: to think that my old compañero is closed off from such beauty! At such times the living room smells of submarines and death. Beats me.

Q: What do you think to be Zoshi's strong points?

A: Speaking from the personal level, he is, as well as being a hell of a nice guy for a slant, a Japanese sandman. While some astute connoisseurs have called him a good, safe, common-sensical, and impeccable mediocrity, to me he has always been wrapped in the robes of the giant! How painfully clear! In terms of art, before which all must stand atremble, for without it can life be worth the candle, or the game the victory? He is, like they say, equally very damn fine and pretty damn interesting in his very lovely care. In his best work, blue milk bathes the beaches. But best of all perhaps is his snappy white coat and Singapore Slings. No crab he, but the smiling face of the lotus and those funny little gardens with the rocks and sand. Known as Jig Jolly in his native San Francisco, yet Higgins ink throbs in his tiny veins.

Q: Earlier you said something about Jung . . . do you feel that he has no place at all in an artist's scheme of things?

A: May the heavens forfend me if I breathed it! I don't comprehend the innovations of these old geezers, but I will defend to the death their right to a nice cushy practice. What I know from Jung you could put in a knothole. I've read somewhere that he once got knocked unconscious dancing the "Clarinet Polka." Take it from there, young one. If it's box-office, why knock it? Freud behooves as one who locked himself up with crazy ladies all day—is it any wonder that Mrs. Freud went around in an old *shmatte*, as biographers imply? Adler strikes as a true crank. Inventing a shoe to make psychopaths feel taller for a few hours, give or take, is not exactly the mark of your huge intellect. But he was a friendly little fella who liked nothing better than to seat himself at a groaning broad.

Q: For you, as a collector, does astrology have any use?

A: My wife has imparted some, sweet girl. Do you know that she is happiest in simple riot behind yonder sand dunes? Often she finds herself in a lubricious act, but it is a long road that does not cross. One pretends gaiety while the vitals throb, *c'est la vie*. I made one of my finest buys, crude word, the Chancré "Boiling Earl," because of what the stars did whisper low. While many scoff or weep all snots and tears, I am the Ty Cobb of the Big Board! At night, they're big and bright, and like Prince Rupert's Drop, hold the secrets of the universe! Saturn in the Summer House spake, so to spake, low. Ere long I implored my spouse who mumbled through her avocado face cream and a mush fulla guava pie that I should send in top bid for Chancré's effort. O fair interpreter of lore arcane! Unto this day I bend the wonted knee to her, the cuddly baggage. You cannot make your sun stand still if he has the runs.

Q: Do you find much to laugh at in the art world today?

A: One must be prepared to proffer a solemn countenance. In the morning I am covered with hair, but I do not cease a jot. Originally discerning gentlemen like myself find it ruth to chortle at a potential million. With paint on their shoes and dopey ideas yet are these daubers possible flushes. I abhor and blench at the word "plight." If it were not for those with ready smiles and a checkbook to hand they would all be no better than what a recent plumber with fedora to match called a "crabwoman mooning lovesick with colors." If you can catch that, put it in a jar. Often I feel an awkward distance in my own occasion from that which is clearly the possibility of something. At such times, I not only feel it meet to stifle my rising chuckle, I do so. If they would only stop writing poems drunk on Ocean Avenue and wielding brushes while grunting, the cause of all this levity would be sorely nipped. Not that I am Ben Grouch, uh-uh, on the contrary, I feel myself to be Felix Randall, "the giggler." The brush, however, does not fall far from the canvas and it's a cheaper mousetrap we all wish to pick up for a song. What happens is more to the point than what doesn't and I'll put on my face the mask of saturnine gloom to wrest from a needy artist an item that may ennoble and long enhance not to mention fetch a few rubles. Any clot of common mind could see that a misplaced guffaw could euchre the works. That is far different, you will agree, from whistling "Dixie."

Q: You feel then, that contemporary art must deal with life as it is lived in all its grime and squalor, not to mention its ineffable lameness?

A: You have hit it right on the brass tacks, my friend. In the truest sense, gloom is all. Think on those little mouths nibbling the goodies from Papa Mondrian's table. I don't trust a man who wears a tie and is always so clean. Wasn't it Lewis Canto who said that it is impossible to be well groomed *all* the time? In his spectacularly thrilling mode of the shabby, who should know better? It shows in his poems to great advantage, you may lay to it! Where is the *poetry* in the hilarious? The *prolific*? No, it is in what you so wittily term "lameness" that we must put our faith as such. Take Fred Fella—a simplike appraisal of his last show might have it that the reoccurring relationships of triangular forms bespeak the, as it were, cheap joke. One whispers "nay"! While there is a quiet, intense wit and care present in each and every manufacture from the brush of this gifted collegian *manqué*, beneath the sheer "brushed" expertise shown to such good advantage in the immaculate galleries of which I have, let us say, a small piece of the action, there is a profound sense of the crippled. Thus the terrible power of the postmodern, than which there is nothing more Malamudesque, or Malamudian, if you prefer. I'll stick with Malamudaisian.

Q: Your command of the critical vocabulary is stunning. I must admit that I was not quite prepared for your subtleties . . .

A: I reply to your transparent flattery with a Yiddish saying taught me on Rivington Street by a gravestone chiseler. "If you put your hand in your pocket you're liable to find nothing." There may be wisdom in this, but how can I ever assume that it will come to this or that substance? I cannot, of course, since our time has a warped view of surrealism. This doesn't bother a sport like me, with other fish to fry and a host of irons in the fire I myself started, but when it is counterpointed and enriched by superb and Daumier-like caricatures, I tend to shirk and even walk swiftly toward the hills. I'm afraid that I don't have it in me anymore to slap my chest and sing "tarantara!" By any measure that I can devise, artistic betters see again that ancient struggle William Blake described. Unfortunately it slips my mind at the moment, but you know Blake. Did you know, by the way, that his great novel of adventure, *Gorgonzoola,* was written while operating a punch press on Pearl Street? It doesn't matter but it's one of those little nuggets of info that make life inestimably richer, as when the moonlight sets a million fairies sparkling among the scrub in Vermont. Yes, I've taken a number of swings at artists who have spurned my generosity and fake enthusiasm, but it is always against them in myself that I've struggled to see a world again alive with unrestricted trade. Often a tear jumps to the eye, or both eyes, when it is once again made clear that the city is a crab of rock. To emerge from this massy maze takes a smidgen of guile and a vocabulary to knock your hat off. Thus am I before you, Miltonian man, a little the worse for wear and tear, but the banks don't cry when I enter, if you follow the parable.

Q: Some years ago, you made a rather cryptic remark, "When in the mountains, paint boats." Can you expand on that?

A: Look, I'm the most happy fella. Cryptic is as cryptic does. Socio-historical and mythical states are few in number, but quite real. Take Wyoming. A boat is a marvelous "thing" because it has no true vortex, not in the way the universities speak of it. That, I think, is prime experience for any painter, no matter his essential numbness. How boring it all seems now, yet we dare not forget! One must remove oneself from one's varied obsessions so that one can truly discover that revelation that may come to one in the long stretches of the night. I am alone but like it that way. If you glance sideways at *Ulysses* on the shelf, its bed of pain, so to speak, from its foxed pages there comes the most eerie green glow. That proves *something.* Who alive with chest of steel within which burns a heart that might, who knows? release a poem some fine day dares say no to this one immutable fact: that it was the blind fop, Aloysius St. James, that very crown of cunning, who tapped the green with his cane and loved to listen to the lions roar? It is from such dandyism that the spirit of the maritime will deliver us.

Q. You think, then, that painting has become . . . precious?

A: All art is precious, simple one. But my eye descries your drift. I despise an art that perpetuates itself. Eat that and have another. In the vasty deeps of midnight dreary, your colorful Matisse may as well be Benny One-Ball, I speak of the imposing canvas that great heart executed while doing a nickel in Dannemora. But "precious"? Pardon my retreat into immoderate laughter, but you are coming off as a dope. Is the echo of greatness enough? I think not. We aghast bystanders and sincere appreciators, busy serving buffet suppers to regiments of cretins in beards and faded blue chambray shirts, want the tender arms of greatness to enfold us, but at night crabs sit with us at bars. How then can we know, though our thoughts be pure as Protestant vanilla, who is which? Out of sere days and days and days of such despair do you wonder that we often seem mere philistines? Though personally I would like to follow those whose work has touched upon the light or darkness, vibrant souls who share the gift of poetry, yet am I constrained to do things of a grossness. I refer, of course, to my recent purchase of West Broadway.

Q: I had meant to ask you about that.

A: While I do not like to defend myself against New York Choctaw, a word may be in order, e'en though many who despise my rakish flair may accuse me of buttering my own biscuit. While Anderson Hollöw, the celebrated critic and translator of "Rock of Ages" from the Swedish in 491 variations, has called this altruistic measure on my part the act of a "comicstrip Launcelot," it must be borne in mind that he prefers to set up more and merrier academies and counteracademies. The constant buzz and murmur of gossip in chic salon and *soirée intime* points with candor to a singular prime fact, namely, that Hollöw is not the embracer of objectivity. In short, though he pretends to be the Mickey Finn of aestheticians, he is a slovenly liberal. It is that what I feel, in the world, is the one thing I know myself not to be, for that instant. You may well ask! The "West Broadway Buy," as it has come to be known among those who treasure well-built Victorian greybrick churches, seems, *seems,* mind you, to have a lovely randomness about it, but actually it is just one of a quietly didactic sequence of proposals. It's jake by me to call a tub a tub! Many have importuned Hollöw to offer nothing but his sprightliest, but he is bent on dumbness.

Q: Do you feel that art is about to make another significant breakthrough?

A: Cultured Brahmins that we are or bend the knee in tearful orison to be, yet how we wish that we were still those mere shells of solid brick and soft sentimental mud. It is, of course, but youth remembered, a cig and a Coke, but the wild singing and general traipsing around! It's like to bust you up. A breakthrough is like an evening. There's no way to start it out directly, first the sun must go down, and Old Sol has his own mind. I mean to say a true breakthrough claims great antiquity. You don't go boo-boo-boo and hot dog! and of a sudden you're dealing with a Cézanne or a Jocko Conlan, no sir. The ways of innovation are strait and fraught with maladroit. Would the French tongue have been reified had the redoubtable Sade not been nabbed for a youthful indiscretion? *Sic transit semper tyrannus,* or as a wag has it, "Hats off! The flag is passing by." And may God help you if you don't whip your bowler from your sconce in such circumstance! One waits for the brave spirit who will paint the inner meaning of the nosepicker who stands bemused before the drum majorette's teeny skirt and panties to match amid a storm of sour fifes.

Q: What of the charge that you tend to buy, not individual paintings and sculptures, but whole studios indiscriminately?

A: To this I turn a half-swollen cheek toward the door. It is only after the cheek, or the door, leaves do I feel this strange joy. More specifically, most men of a throb of sleep would be bored to sickness walking through all this caca! I tell you, strength is not often drawn from the doors of hotels in the darkest quarter of the city, no matter the protestations of those who shave their legs and pubic hair for love. Sheets of delirious colors pall. The unit is the thing, so modern, so *dernier cri,* so awash with candor. A man in his right mind, and financially well fixed, such as your devoted partner in colloquy, should stand still for a whole room full of concepts, sand, rocks, and ropes? Consider how explicit the activity of light is. Explicit and methodically engineered to bore you out of your very trousers! One affects a passionate interest, not to mention an overweening curiosity, but basic human relationships flower when the sound of crisp lettuce is heard in the land. The apparent melding of a vocabulary involved with symbolic action and other phenomena is duck soup say once a year. More than that, you could get a slight twinge of nausea and get so crazy that you might pity old clouds devoured by clouds of hot sand. No telling what! To protect myself from sinus headache and general waves of *mal de mer* brought on by twangs, drawls, and memories of peanut butter pie, I break out the checkbook and the old Paper-Mate and take everything off their hands. It may be, as you suggest, done "indiscriminately," but no one can accuse me of doing it with love. The sweet cats rarely complain, rather is a rose bowl held in the hand more their speed. I am a sort of shining champ, titter who will!

Q: Is that a loud crashing coming from the gazebo?

A: My helpmate and the faithful Shreve aromp. It is nothing more or less than a throe of joy. I grow weary of the problem of historicity. Though what interests me goes on all the time an occasional boredom afflicts continually. They don't have weather in Ohio. Look, I cannot abide to be predicated. That seems painfully clear. My not inconsiderable wealth allows me such activity, if such it be. One might call it an instance of the archetypal nature of it all, around and around. Though what constitutes a true archetypal nature—and they are, like a good man, hard to find—is as nebulous as the idea of admitting the fact of one's own feelings.

Q: Is there such a thing as the artist's "plight"?

A: To think on the rusty metal and the paint-spattered shoes may be enough for some. People enjoy being covered with soot, *de gustibus*. Think on the Romantics. Think on the watery principles of Locke, if you dare! Men consider that heat damps the money in their pockets but do you hear a peep from them? Maybe some quixotic mouthful like "forget the money," or "forget the desperate stretch," or "see the passion of art." In the meantime rumors are flying. This is why the ferns are exciting. Look, the wings of disbelief and beauty may be what you incarnate as the "plight" of the artist. I don't cotton to it! Despite flowered shirts, rawhide vests, suede ties and the like, a casual glance will reveal a plethora of perfectly excellent hucksters. The butter, like they say, would not melt in their mouths. In the green of back Brooklyn, if you follow my argument, there are many cantaloupes, equally some very damn fine splendor—I mean despite the anguished faces and other symbols of decrepitude in service to the Muse. Breathes there a sensitive with soul so dead that he feels no pinch of ecstasy at being told that he is untransmittable? Yet withal this is a clumsy blague. A large moiety of these starstruck folk bring to mind an array of brand-new decks of Bicycles. As a mere fringe figure huddled on the edge of this great volcano belching fashion I keep shut however, with the exception of a huge and spectacularly insincere smile, the trap. So you see, don't talk to me from the "plight."

Q: I know it's getting late, sir, so allow me to ask you a final question. What do you think your direction as a collector and connoisseur will be in the years to come?

A: Love so seen in its place is always there, if you will permit a touch of harmless folderol. I see myself at times curving back into the mammoth pool, yet at other times, equally insistent, my bones turn to dark emeralds. It's a problem, these thoughts we have not yet thought. "Direction" to a man like me is more than just your going down the block to the Bijou or the other way to pick up a bottle of booze. I delight in the *sense* of direction all by itself, almost as if it were a large white flower or a pear, rotten maybe, but so complacently *there*. That's the kind of hairpin I am. Some say that the image of going, or of movement, is clear as the eye of a chicken, but I disagree. In my day I've taken the garbage out and I've taken the saw back to the garage too. Little would you think so to look on my bronzed face and so on, but I too have been humble. Now, I think of happiness as that warm center, maybe with the flesh caving in? One thinks of Rilke. His remark to Wasserman is of the quintessence here: "I don't know where you live but I'm going there." Such stilled beauty may be a kind of lethal compromise with the *Zeitgeist*, but can anyone refute such genuine feeling? I look for that lost and gentle farmer whose scarlet face betokens the true lush. It is there, I feel, that Kafka's castle must be built. Direction is a winning form of rhetoric and we must look to it for new aesthetic structures. Huddled cribs, tiptoeing wind, and squinting leaves—toward them I set my rather handsome face. And why not?

CLIMBING

ROBERT MORGAN

Say the hillside pasture is the foot
of a long ramp running up to the first
tier of the ziggurat
and the ridge beyond steepens
up to another
shelf and the mountain
glides up to a table,
and nothing beyond that but
blue chimneys sharp against it.
Think how far that nothing goes and how wide.
Its other side touches the Other
Side. Though it's just three
big steps down to here
where plowed ground and cropped
grass are separated by a fence.

Rising a step at a time out of the valley—
finding stirrups in the dirt
to swing up on,
choosing
the next in mid-step
with no pause, one foot
rolling onto the next. Fingers
root for a hold in moss,
leaves, the dirt at your face.
And always the top just above with
the eastern sky behind pure,
almost black in the afternoon.
Once up rest

as in a bunk or hayloft
against the skyeave and look down.
Nothing to do but go down
and you go,
banking on rocks
and rappelling off saplings,
lose altitude so fast ears close,
dropping, parachuting,
till the ground holds out
on all sides,
thickets and marshes to crawl through.

At the timberline trees compact
and twist to hold in their sap against emptiness.
Forests give way to thickets that give
way to arctic moss and above just
weather busting rock on the pile.
From here the climb's rough
as the phrenology of mountains to look out on.
Lofty shore I climb out of the deep
spruce and rest on a shelf in the heather.
Foamy heather in the sun.
Tatterdemalion coastline. Nothing
ahead but clouds breaking spray on the turret
rocks. Let me camp here in the surf shrubs,
near the island's polar coign.

Sometimes I feel the seethe and crackle
of ions swept off the sun
as solar wind,
brushing the heat wool
and spiraling up blizzards
that far out cool and fall, each
particle refinding
its mate in the platinum mud.
As a tree sends out limbs grubbing
and dirt's hungry for sun milk
through rot and leafmold,
mind too wants the slimy core
metals, whey and stink
of the smelter.

Wants the steel honey
of the blast furnace.
Eye climbs open and finds the sun
high over the river, the clear stuff
drifting all the way down to the sky sill.

Beans want to climb.
They lift themselves up
and feel around for something to hold to,
runners already kinked, constricting
on string or cornstalks, felted
to catch any surface.
But cucumbers prefer to spill
out of the ground and run down
hill holding leaves overhead.
Have the same inclination
as water,
to pour and keep going.
Aspire only to describe the terrain,
seek no skeleton.
So you train daily,
twisting the woolly runners on hemp.
Lifting and tying under armpits,
propping
so they stand
toward the azure they don't want.
And at the top fall over again, streaming
out kitetails, hunting the ground.

To live in the mountains high
as in an attic with
dormer windows
and balconies,
castellated ledges
looking out over the plain
inked green by the little
lost river hunting
a sandtrap to vanish in.
The cornfields and sheep are
descended to by
fitting feet and hands

like cogteeth in holes up and
down the cliff's sheer.
Cool in the rock,
wind this high
always freshing the recesses
and playing the tunneled
passages where
grain is heaped and
cisterns of water cool
under tribal paintings.
Images gathered like honey
and brought here. Cliffdwelling.

In the still
of early morning
smoke climbs
scandent
taking hold and
lifting high
on the coolness;
a few shifts
and twists
out of true
alignment but
raises like
a charmed snake
from the house
clinging
acrobatically
to altitude,
and jacking almost
straight plumbs
the upper air
still reaching,
touches
a current
that spreads it
over the valley.

The two big pines that planted the grove
below stood among hardwoods.

Their shade was a dank
yard, a briary
tent, drifted with tufts
of needles from the heights.
The voices in the towers became
an obsession.
Looking up one of the masts
radiating its spokes you
saw no further
than the first branches.
Limbs near the ground had fallen
but the stubs remained in the bark
or cores of the stubs pegging
the battlement-sized base.
Climb up them to the first limbs (a pop
warning how much weight to trust).
From there on it's mounting
a spiral ladder, brushing
aside to find the next hold.
A dry limb breaks
numbing the hand, sickens
bones. The trunk becomes
tree-sized, green leather bark.
A squirrel's nest stuffed in the forks.
Resined hands grip anything.
Already above the hardwoods, looking
out windows in the branches.
Swaying now up into
the christmastree top
and easing into a saddle of limbs
just under the tip. Body weight
makes the sway longer, like a metronome,
going far out over the other trees
and back, canter holding
the bristly reins
looking over the pines
to the pasture. The steed takes off
as wind returns
spilling its voice
around and below you.

VICTOR

An excerpt from the novel *Island People*

COLEMAN DOWELL

"It's open, Victor. Be with you pronto." There was a pause, then "O.K." came at him from near the floor where he knew the young man was crouching, exchanging caresses with Miss Gold; their pleasure in each other had given Chris an answer that seemed both satisfactory and tactful to one of Victor's disquieting questions. During the Puerto Rican's second visit to the apartment, possibly careless because of the vodka and orange juice he had drunk in large quantity, he asked Chris what made him feel so sure that he would not get jumped and tied up, made helpless to stop Victor from making off with, as he put it vaguely, "things." The dachshund bitch was asleep with her head on Victor's leg when he asked the question; Chris waited until the other's challenging eyes met his before he gave the answer which consisted of a nod toward the bitch and a word, "Her."

"You trust her, huh." On his tongue the word 'trust' sounded more like thrust. Chris agreed that he did.

"But not me so much, huh Chris." His smile was lazy; the words were lazy too, as if they were the smallest small talk. "You don' thrust me so much."

Chris answered him in a bantering-serious tone he had found to be most effective with Puerto Ricans.

"She trusts you, I trust her. O.K.?"

"O.K., Chris. Shake, *amigo*." The allusions had continued, however, worse for being indirect, and today Chris was determined—despite a tactical error made earlier in the day—to bring things into the open. Under the circumstances, the 'things' he meant to expose were as vague as those Victor had meant to steal: neither man could be sure what the things were until they had taken open inventory and some sense or desire told them *this is it*.

Standing in the living room out of sight of his visitor, giving to them both a moment's more grace, Chris had to admit that the past two hours of analysis of himself and the boy and the progression of their relationship had opened no unopened door. He still stood with the boy in a corridor of many doors, most of them tight shut as they had been before he started the exploration. He was unwilling to admit to fear of the boy (or young man; Victor was nineteen or twenty-six, depending upon which of his answers one chose to believe), for that would have demanded severance. It would of course have to come eventually, for what on earth, in any long-range plan of his life, could he do with an unemployed Puerto Rican erstwhile delivery boy? So he let *disquiet* serve as the word for what he felt in the charged air between him and his unseen crouching guest, knowing that it could be transformed by word or gesture or even prolonged thought into what lay beyond and above disquiet—visualizing the progression as stair steps with name plates—one place below horror.

His desire to deny fear a name plate (disquiet was permissible, horror improbable) and to re-establish the old footing where he was the more secure of the two, brought him close to blurting out a question that would have put him at a further disadvantage by causing the other's defensiveness to surface: "Why did you come an hour early, without at least telephoning?"

He had had the doorman send the boy away. Victor had come at eleven and Chris had told the doorman to tell him to come back at one instead of twelve, the usual time. But still the need to know why the rule had been flouted rankled Chris, so that he prolonged the silence, adding his sharp displeasure to the charged air that separated them with increasing incisiveness, as if it were assuming the shape of a knife-blade which could sever whatever their bond had been.

The thought of the severance of the bond before he was ready to will it, and, far from the least consideration, what asking the question would reveal about him to the boy, caused him to break the silence by going into the kitchen and rattling glasses and ice until he could trust his voice not to give him away. He had learned to fake, for the benefit of these people, an insouciance not native to him. There had been times in the past when, beset by worries, he had longed to do the unburdening, but what he thought to be an accurate sense of his role had restrained him. Holding back his own emotions had, paradoxically, a draining effect upon him. Sometimes, when they had left him, he felt like an empty husk stretched out on the couch in his study. The facts of their lives—heavy, ugly, valuable to him and his writing—hung somewhere in the room above him like tatters of dehydrated meat, as weightless at those times as he.

He listened to the silence coming in waves from the floor of the entrance-way. He imagined the two creatures crouching there, staring in his direction, both gifted with abnormal instinct.

"Drink, Victor?" He counted the beats.

"Sure."

"Hang your jacket in the closet and go on back. I'll be right there."

He listened in vain for the rattle of coat hangers, but in a moment he heard the squeak of Victor's sneakers on the tiles as he went to the study, apparently carrying Miss Gold as there was no accompanying clack of her claws. Victor's tenderness to the animal reassured him again, and he mixed Victor's drink and poured orange juice for himself. He put the drinks on a tray and turned on the kitchen radio. He had found that distant music—or sound; commercials would do—encouraged the confidences of people who in their own neighborhoods inhabited a continuous stream of sounds composed of rhythms and melodies from an almost infinite variety of sources: radio, TV sets, mouths, hands beating bongo drums and garbage cans, sticks on metal railings, feet clicking sharp-toed high-heeled shoes on stoops and pavements. To drive through El Barrio, as Chris often did under cover of night, was to submerge oneself in an element of the density and fluidity of sea water, beneath which one, as an alien substance, sank, but upon the surface of which the buoyant natives floated. Chris imagined that if all possible means of creating sound were suddenly cut off, the Puerto Ricans would plummet to the dry bed of their lives and die, blind, deaf, and airless. Sound was their element, and in his apartment he provided it, but on his own terms: distantly.

He let himself into the hall under cover of a nattering commercial for a hair color.

"Hey, Chris, what's this ting here."

Chris had carefully made no sound; glasses, tray, door had all been swathed in his calculation. Notes appeared, ghost-written, in his mind: *Highly developed sensory perceptions—instinct for survival—* "What?" he shouted, creating distance.

"This-a ting. What's it?"

"Minute, Victor. I can't hear you." He opened the kitchen door, stepped partly inside, made an exit, slamming the door on a tide of words.

"Damn the amplified human voice, damn commercials, damn radio. For that matter, damn Marchese Guglielmo Marconi—" Experimentally he stopped where he had stood before. "O.K. now, shoot."

"Hey, man, you know I give that up."

"*¿Que dice—*" he began, amended it to: "*¿Que tu dijistes?*"

"Needle—shootin'—*comprende?* Aw hell, joke."

"*Dios quera que tu estes ralajando.*"

"*¿Que?*"

"Not a thing. *¿Qual fue la prequenta?*"

"I said what the hell's this ting in here—buttons, mirrors—"

"Oh." He spoke carefully over the music. "In my closet. Don't shut Miss Gold in there, please."

"I ask a question, Chris. *Is this goddamn thing a microphone?*" Chris stepped through the doorway quietly. "That's a sunlamp, Victor, the poor-but-honest man's rebuttal to the pejorative statement of March . . . *Märchen?* No, I guess it doesn't work, after all."

"It don' work. Sssssss." He seemed to be hissing the useless sunlamp. He closed the closet door with a bang, startling Miss Gold. He knelt, contrite, and fondled her flowing ears, murmuring to her in Puerto Rican, which, in quantity, Chris would never be able to understand.

Chris's smile was balanced; nothing slid or clinked against anything else. There could be no doubt to an observer that a delivery boy going through his personal closet had not disturbed the balance of his self-assurance. As there was no observer he patiently held the smile until Victor chose to glance upward, then, like a teacher explaining the subtlety of a drawing through slight overemphases, he deepened the lines of self-amusement; it was unthinkable that Victor should be allowed to imagine that Chris was amused at him because he did not understand the attempted play on words.

"I meant that my pun didn't work." He set the tray on a table; colon or period? He decided upon a semicolon. "The *sunlamp* works fine. I give you myself as proof."

Victor looked him up and down, rising; his attention was filled with critical earnestness, a quality he brought to bear upon any question, however trivial (Chris imagined), for which his consideration was requested.

"You lookin' good, Chris. Healthy. Sunlamp, huh?" In a series of movements that Chris thought of as peculiar to the Puerto Ricans, Victor crossed his arms over his chest, hands grasping biceps; his feet straddled, toes turned slightly inward, the more comfortably to support the sagging weight of his pelvis which pushed the loins forward. Head cocked to one side, mouth pursed and lightly indented at the corners, he shifted his bright, interested gaze from Chris to the closet door, as though he might be able to see through to the instrument that could flood a March-dark room with tropical sun on command, and his prolonged, quizzically approving nods affected his whole body.

Chris found the performance eloquent and somehow sensual, for his own expenditure of gesture was more carefully considered than that of his bank account. Victor smiled as though he could read Chris's thoughts. Chris fell into a bemused state, habitual though usually unobserved when he was faced with the boy's beauty after a separation.

Victor's teeth were as stainless as good milk. On his beautifully boned, tautly fleshed face, there was no token of former dope addiction. Aware of the eyes grown distantly appraising, Chris continued to gaze as though Victor were the model, he the pupil, in a class of anatomy. He looked at the boy's neck, a perfect, unfluted column of dark marble planted on plateau-wide shoulders like a symbol erected on a mountaintop by a cult of phallus worshippers.

"There," he thought, and turned aside to the drinks, handing Victor his without looking at him.

"Sit down."

"You got a hanger for my jacket?" Chris shot him an unguarded look that asked, 'What did you see in my closet, if not hangers—' Victor shrugged, smiled lightly.

"I didn't want to mess up you good suits. My jacket's wet. I better hang it in the bat'room, huh, Chris?"

Chris gave him a hanger from the closet, seeing for the first time that the jacket was indeed sopping wet, as was Victor's hair, both shiny materials: the jacket nylon, the hair slicked with oil in an effort to make 'good' hair out of it.

Victor hung the jacket from the showerhead in the bathroom that opened off the study. His voice was flat; to Chris's ear, the lack of accusation rang as a deliberate omission meant to heighten the effect: "I was early today because it was rainin' and I didn' have no place to go." He came into the room, face too brightly lit for the words. "No pesos for the movies. I try to sneak in but they catch. Me, Victor Ramos! 'What you name, *boy*.' 'Up yours, *gringo*.' I move up Eighty-six Street, RKO, right? Stand under marquee, lookin at the pichurs, right? Lady in chain' boot' get nervous. Four, fi' *gringos* standin' there but me, Victor Ramos, she get nervous." He clapped Chris on the shoulder. "Right, Chris, huh?"

He sat down beside Chris on the sofa, then jumped up and got a newspaper that he spread carefully on the sofa and sat on, moving his buttocks vigorously. He explained. "My pants wet, man." He shifted back to front, side to side, in a slow circle, watching Chris. "Somebody say to me, 'Man, what's new?'" He hopped up and bent over. "I say, 'Man, read my ass.'" He had succeeded in transferring a smudge of black to the seat of his faded khakis. When he remained bent over, Chris indulgently leaned forward as if to read. Softly, straightening and turning, Victor told Chris, "Man, that's not braille." Keeping his eyes on Chris's face, shifting them from spot to spot as though looking for the hole through which the color might leak out, he stooped to pick up Miss Gold, who had edged up to investigate. She snarled and snapped at the slowly descending hands. On Victor's face Chris saw understanding for her actions and admiration for her perceptions (Chris thought, filled with stillness), and a glimmer of what to Chris seemed like surprised recognition, the kind that suddenly revealed adversaries give to each other.

Victor held his crouch, arms dangling, lifting his face, which was on a level with Chris's, and giving him a long, slow look— eyes narrowed, mouth curled at one corner. A trick of Chris's vision in the rainy light caused the face to loom, to become disproportionate to the body dwindling away behind. As if to identify the hallucination for Chris, his tricky vision made the face, from hairline to chin, appear to detach and shift while the ears remained stationary and the glittering eyes achieved an artificial depth; it was the merciless Noh mask of Chris's dream that had brought him starting sweatily awake last night.

Victor bounced onto the couch beside Chris saying gleefully, "Hey, Chris, you gotta watchdog. Ol Pen'house look out for you, man. She don' let no Pota Rican Negro stick his ass in you face!" He laughed with delight and drank, banging the sofa beside him with one hand, a strongly accented rhythm to which Miss Gold responded by standing on her hindlegs, asking to be helped up, her tail lashing in vehement friendship.

Chris began forming a remark, casual and obvious, about the dog's love for Victor, in which he would make the unemphatic point that dogs (and friends) could be forgiven testiness, and even rudeness, if . . . With coldness, listening to his proposed oblique apology for sending Victor away, he heard, as the boy would have done if the words had been said, the half-hidden plea in the 'if.' Loathing skittered upon his flesh like insects. He sought imperviousness through inward arrogance that would turn his dislike upon the boy; finding it, feeling it course through him, exultant to be released, he controlled it, arrogance controlling arrogance, and stopped it short at his skin's surface where it ran spirally in whorls like fingerprints, hundreds of whirlpools of identity, and then ran together in a protective coating of blandness. On the wings of accomplishment he began the interview.

"Victor, old boy, I thought today we might veer a bit from the straight and narrow—ah—narrative, as it were, of indignities, discriminations, deprivations, et cetera—and, with your sanction, of course, sort of—ah—dive headlong into a colloquy, though I imagine that word is inapt, really, as well as undesirable for our purposes; you do agree that formality, even as a grace note, would be more of a hindrance than a help?" He bent upon Victor the parodied look of an equal and noticed, with a slight inward sinking, that Victor's outward blandness matched his own.

Victor delicately flicked the ash from his cigarette toward, but not into, the ash tray.

"Boil it down for me, Chris," he said and stretched his legs to the coffee table. The words and gestures conjured perfectly a Madison Avenue conference room, or at least the television version familiar to them both. After his acknowledgment of the boy's power of mimicry, he thought with peculiar relief that it was really only mimicry of an impersonation which children, or even monkeys, could do. The thought aided his free-flowing laughter.

When he had finished, unjoined by Victor except for quizzical eyebrows and lifted corner of mouth, he lay back in the cushions in an attitude that he had memorized long ago in college during bull-sessions that he had attended but not joined. It was a position of alert relaxation that all the men but Chris would assume (Chris was always perched somewhere, preferably on the edge of a desk or table), as if at a signal pitched too high for Chris's ear, when the talk was about to turn to sex. He had thought then, as distasteful as procedure and talk were to him, that the series of movements were like the letters of the perfect word to describe what was to come: the falling-apart legs, the down-thrusting spine that jutted the genitals into prominence, the anticipation, which chose to appear half asleep. His own performance was a modified version, informed by characteristic caution and dislike of calling attention to his body, just as Victor's, following the leader, was typically unrestrained. Even Miss Gold surrendered to the mood and stretched on the floor between them, her rump to Victor, her head to Chris. At once her eyes filmed over with trusting sleep.

Some brittleness in Chris melted as he looked at her flowing goldenly on the rug. Often she slept on her back, but that was a gift for him alone. For all her fondness for Victor, she kept her vulnerable belly hidden from him as she slept. With definite longing but with meaning disarranged by a weighty descent of drowsiness as abruptly dismaying as an insult from a friend, Chris thought that if she ever trustingly exposed her soft underside to Victor in sleep then so would he.

He pushed at the need for sleep with the first weapon at hand, words.

"I thought we might just talk today," he said, "about ourselves—" he stifled a yawn and smiled, "—together and apart. Ambitions, accomplishments, frustrations, even, but not—ah—sociological." He felt no longer godlike; he seemed even to lack purpose beyond prolonging physical sensations that seemed as though they were being brought from his past by the generous hand of an old love. The pattering rain, the warm face only a few feet from his own; to accommodate Miss Gold, Victor had slued about until he stretched on a diagonal, his head with its warm secret breath sliding on the cushions nearer Chris. The coolness of the room, the need to draw together, was pointed up by immobility. Chris fell a few inches toward Victor, seeking the heat of his body as unthinkingly as Miss Gold might have done.

In Victor's left eye a line of light appeared, standing vertically from rim to rim of the pupil, a thread of gold edging a barely cracked door. It bellied slightly, became a cat's eye, dark-brown and gold chalcedony flecked with christopher. Christopher lay upon the eye and peered into the widening slit like a fly upon a well of light. "*Laisse-moi plonger dans tes beaux yeux*," he buzzed or droned or sang.

Victor cleared his throat, and Chris lay upon a flat lightless surface.

"What you wan' to ask me, Chris?" Chris sat up.

"No, no," he said, hearing rather than feeling the testiness. "It's not that I want to *ask* you anything." Victor spread his hands. Chris tried briskness. "It occurred to me that we've never had a conversation. We've had interviews, monologues, confessions—" He spoke the last word with distaste and covered it with amplification as Victor muttered, "I don' confess nothin—" "You telling me about the—ah—heroin was a confession, because it was not something I knew about or suspected," though this was not true. Still, Victor's face cleared, and he nodded. Chris went on, "A conversation can, but needn't, be confessional. What I want for us is simply to talk." He believed himself to be recovering nicely. "As friends." He found that he could not emphasize the last word.

"*Amigos*," Victor said earnestly. To Chris the word seemed suddenly too warm, to be redolent of chili powder, and chocolate misused in sauces.

"Friends," he said, and then nodded as if to counter the meant refutation. Victor grinned; it seemed to Chris that the grin had an edge of satirical implication. Victor nodded vigorously.

"O.K., Chris, you start." Chris went as dry as crumbs on a plate. He felt like saying, 'No, no,' again; 'No, no; you don't do it that way,' like Miss Havisham telling Pip to *play*. But what was conversation, in the context of himself and Victor? He thought with hardness that the diplomatic meaning was the only one that could apply: representatives of two (opposing) countries exchanging hard-nosed policies in the guise of informality. He had been an idiot to imagine anything else was possible.

"Very well—O.K." he said, smiling thinly. "Aims, ambitions, that would be a topic. Or accomplishments. Do you want to choose, Mr. Ambassador?"

"Yeah, O.K. How about—ah—accomplishments?" Chris had the unpleasant experience of hearing himself as others heard him, the 'others' for whom Victor undoubtedly gave a fuller impersonation. He had learned how clever Victor could be, at showing you a thing you had not seen before. In that particular case it had been Victor's aping of the old spastic junkie who had turned him on to heroin: knee bent, pigeon-toed, he had flapped around the room with such realism that Chris had smelled the stench of the bottomless evil that such a man must give off. When, realist, he had asked how such a man could administer the needle to a novice, Victor said, "Oh, that *hombre*. His han' steady like a rock holdin' a needle. All this stuff—" he staggered around again—"I dunno. To make you sorrow?"

In a bright flash connected with the implied con game, Chris saw a thing that for all his nosing about it all day he had not seen before: Victor came to him for one reason only, which was the ten dollars he was paid for each 'interview.' He had come early today because he was broke. Simultaneous with the flash, a practical Chris was asking himself why else the boy should be expected to come.

Expressionless, Chris reached into the pocket of his jacket folded on the arm of the sofa and drew out his wallet. As slowly as he dared, deliberate to the edge of insult, he isolated a ten-dollar bill from the company of twenties and fives and turned full-face to Victor, expecting some version of the usual byplay: always at this point Victor would have his attention focused somewhere else— upon a book he had taken from the shelves and pretended to read, or, most often, upon Miss Gold, with whom he would instigate a sudden floor-sprawling tussle, realistic in its passion. Once he had bitten her stomach and she had screamed like a human.

Victor had put his glass on the table and bent to Miss Gold, his hands rousing her in what Chris saw as a purely reflexive action, for although his head was bent to her, his eyes, wide open and staring with strain, were cut upon Chris's hand and the money in it. Chris rubbed the bill between forefinger and thumb. A vein throbbed in Victor's temple. The money whispered in Chris's fingers, responding to his caress. As though the bill were a small unimaginably depraved creature that threatened all morality with suggestive whispers, Chris closed his hand upon it with audible suddenness, or so it seemed to him. He would have sworn that the strangled creature gave a cry that bloomed in the room like an articulate lily. Behind Victor's head, silhouetting it in kinky-haired blackness, was the white flash of the lily's blooming, and the room was streaked with layers of odor as though each door concealed a corpse with a lily growing from it.

Through veils of hallucination, Chris saw that Victor sat stonily gazing at the corner of green pushing from his closed fist.

Victor's head swiveled around, and his eyes rested on Miss Gold, whose chest his right hand cupped. He drew the hand from her chest to her side, ran it up and over her shoulder, along the side of her head, his fingers flicking her ear so that it folded back on itself like a calyx, a ridge of supportive cartilage holding it like the rim of a cup above and around the exposed, sculptured heart. He leaned toward her so slowly, in an echo of Chris's slow motion handling of the money, that Chris could see each separate muscular transaction that disfigured the face until it was unrecognizable in its cruelty. The skin around the eyes tightened until the eyes lengthened and lay flat; the nostrils flared like those of a terrified horse, and from their outer edges deep-cut lines ran to the corners of the mouth, which, lip lifted, showed grinding teeth. At the edge of each jawbone a round knob stood out and pulsed, with an indentation in the centers that looked as if a thumb had been pressed into balls of clay.

Miss Gold watched unmoving except for the gradual lifting of hackles. Chris's hand moved slightly to the left of him until his fingers clung to a surface and then inched inward as though expecting some encounter. At that moment Victor opened his mouth and a great, doglike noise jumped from his throat. In her haste to flee, Miss Gold fell onto her back and lay with threshing feet while Victor roared with pleasure. He fell forward upon her, his knees hitting the rug on each side of her swollen belly with no room to spare. The room grew dim for Chris, and there was a pounding at the base of his skull.

"Gobble, gobble," said Victor, bending over so that Miss Gold's wildly flicking tongue could graze his chin. He got up grinning, his high spirits as shiny as a child's.

"Hey, man, I'm *hongry,* man," and Chris got up, thinking dimly of violence as an appetite-sharpener.

Victor flung his arm high around Chris so that his armpit was a socket for the rounded bone of Chris's shoulder, a perfect fit. Chris was repelled and excited by the union of sweaty, acridly virile cavity and linen-clad shoulder faintly scented with the sandalwood he kept among his shirts. They walked down the long hall that way, uniquely joined Siamese twins. Chris could feel at the jointure a pulsing that belonged to both of them, as though at that meeting of their thinly veiled flesh was where the beat of their life was concentrated, a mutual heart.

At the door to the kitchen Victor released him. Chris opened the door and gestured the boy through, then secretly felt of his shoulder to see what of himself Victor had left there. Feeling the shirt slightly damp, he had a moment of elation, thinking 'blood brothers.' Jimmy Weldon stood in his mind, asking Chris to join him in the ritual. It had been a perfect moment, perfectly destroyed then, as now it was, by the memory of refusal. Chris had longed for the fusion but fear of being cut had made him turn away.

He took a long knife from a drawer, Victor pantomiming terror, and opened the refrigerator saying, "There's some kind of loaf here, my housekeeper made it yesterday. Chicken, I think, and veal. Pork, too, probably. I had a sliver." His impulse was right— 'some kind of loaf,' but he could not stop himself from giving its name and fame, wondering as he spoke if his purpose was to intimidate Victor. "*Pâté en croute, Bourguignonne.* It took her most of the day. The puff paste, the crust, has a thousand layers." He set it down on the counter in front of Victor, who started back from it, dramatically dazzled by its fame. It was a rebuke well delivered, and Chris laughed and relaxed, saying, "Here, help yourself. Or shall I . . . ?" Was there an implication that Victor wouldn't know how to serve himself? Victor nodded. Chris brought the knife down, indicating a generous slice. "About like this?"

Victor spoke without looking at Chris. "You still dietin', huh."

"Same old liquid diet. Vodka, mostly." He managed to conceal the fact that he was not drinking during the sessions by having orange juice, an improvement over the strong cold tea he had had to drink in quantity to mislead the bourbon drinkers of past investigations, during his Negroid phase.

"Imagine that, man."

"Like this, Victor?"

"All."

"All—?"

"The whole damn loaf. I don' need no damn diet."

Chris gave the knife silently, watched the other's fingers close familiarly around the handle. The fingers knew the handle, he thought in deep bemusement, the way his own fingers knew a pen or a typewriter. Thinking 'What is he waiting for,' he heard with curious anger Miss Gold's scratch at the closed door. He turned to open it for her, seeing, as he lifted his eyes from Victor's hand to his face—fleetingly, as he turned—that Victor wore a look of heightened stillness.

Opening the door, Chris saw the afterimage facing him as if it were Victor entering rather than the bitch; he was thus able to verify the fleeting impression that stillness is without color, is only a pale hue, for Victor's organically vibrant bronze had faded until he was the same shade as Chris, a vague pinky (sunlamp) whitish-grey, the monotone of ashes.

Miss Gold sat up politely at Victor's feet, her paws kneading the air until she found her balance.

"You," Victor addressed her. "You don' get no chicken-veal-pork-t'ousand-layer-somekinda-loaf. You too fat, Pen'house." He cut a two-inch slice of the loaf and balanced it on his palm. With his toe he nudged Miss Gold's stomach. "Pen'house too fat, *amigo*, you know?"

"I know. The vet put her on a diet yesterday."

Victor gave Chris a strange look, his face puckered up as though he would burst into tears. Chris gazed at him, baffled. Here, in this look, he thought, formulating a thesis as he went along—here in the boy's reaction to the simple remark, which was a commonplace between dog-owners on the elevators Chris rode and on the side streets he haunted in company with the similarly happily servile, was contained the kernel of difference, potentially powerful as an acre of bombs, which separated—more than money, morals, education, the lot—'the classes.' Watching Victor turn his frightened look on Miss Gold, seeing the look become a compound of brutality and envy, he thought that it was as if the two were contenders for the same bone. That's it, he thought, with the thrill of discovery: the difference is in our attitudes toward animals—the way in which an animal is seen as an equal: as companion or rival!

As though to bear him out, Victor said, "Dogs. Dogs, even," and tilted his hand, letting the slice of meat and crust fall to the floor. Miss Gold had it almost before it landed.

Chris walked abruptly from the kitchen into the many-windowed dining room. He glanced at the rain-pocked reservoir, seething with frustration. Something had gone awry in the past moments, somehow had become obscured, which the dropping of the pâté to the floor had almost recalled, but not quite, like a mechanism set up to trigger memory but which turned out to be faulty in execution.

"Delivery on the way." The doorman, an old man who calculated his privileges as he calculated his tips, upon an assumption of the degree of a sense of *noblesse oblige* in the person to whom he extended his hand at Christmastime or spoke to throughout the year, was terse almost to insolence, as usual when he spoke to Chris. His were the bad manners of the old family retainer down home (white; Negroes in Chris's native South had perfected a much more complex manner of asserting privilege which contained almost no benevolence while seeming to be composed of it entirely), to which was added the special New York ingredient of impudence unalterable by the years. Chris and this particular doorman did not like each other at all, but understood each other in the matter of attitudes, which included the understanding that the doorman's deference was saved for the dispensers of too-large tips, who provided his luxuries and for whom he felt contempt.

"No delivery expected," Chris said, patient and distant as if the old man were the gardener's idiot child. "I haven't ordered anything." He took the receiver from his ear, intending to hang up before the doorman could beat him to it, an old game. The doorman's voice squawked demandingly in the air—"Ramos"—

Chris saw that the old man had tricked him again, deliberately putting him in the position of having to break his rule never to see anyone before twelve noon, or else making him side with the doorman by giving the old man the satisfaction of sending Victor away, unreprimanded for pretending not to recognize the boy. Chris's repeated instructions to him to treat the boy as he would any other guest provided a constant challenge. The old man's content at the predicament came through the loudspeaker green as laurel. His patient waiting for Chris's voice was that of a hero lacking only the weight of the crown upon his brow, which even now was suspended above him crowning him truly with its foreshadow that was like the spirit of his heroic act, itself a foreshadowing of the laurel which in turn was its shadow. Thus did Christ try through intricacy amounting to reductions to naught to calm himself and regain perspective.

The old man's little victories could not be allowed to count, but his, Chris's, time *did* count, as did his instructions to Victor, however tactfully indirect the form they had, in the past, taken. It was not a question of whether or not Chris, in the sense of being up and dressed, with effects in order, was free to see Victor or anyone earlier; it was a matter of the discipline of civilities, being impressed upon those in whom there was a deficiency, and upon Victor in particular, for his own good. Chris had thought that he had taught Victor this; his punctuality, until today, had been something in which Chris had taken pride and had assumed that Victor did, too. It had been proof that certain traits thought to be ethnic were nothing more than weeds to be yanked up by the roots by a sympathetic gardener. But the weeds obviously were more deep-rooted than the gardener had imagined. Hearing the doorman's impatient snort in the receiver, Chris told him, visualizing his words as bees entering the old man's ear, "Please ask *Mister* Ramos to come back at one," and he hung up quickly, with a little snap of the receiver on the hook, wondering if the click that preceded his were only in his imagination.

Miss Gold, more seal than dachshund, had taken up a position of waiting by the front door when Chris had been called to the phone. He hurried out to her and picked her up, one hand on her pouter-pigeon chest, the other cupping her bottom, and carried her, seated in his hand, his lips against the bony ridge of her cranium, down the hallway to his study. He placed her on a leather chaise amorphous with sweaters and woolen ski clothes which she had fashioned into a deep cavelike retreat, into which she liked to disappear for hours at a time, until he went looking for her, digging her out, impatient for her warmth in his arms.

But the snug dark held no charms for her now. As Chris watched, annoyed, she threw herself purposefully over the edge of the chaise, grunting when she hit the rug in a splay-legged position, her chest and stomach taking the brunt of the fall. With hardly a pause to assume a reasonable walking stance she took herself from the room with a dignity that should have been ludicrous but was not, to Chris, who took his cues from her. Sometimes she clowned deliberately, and then he laughed, unable not to in view of her considerable artistry. Her dignity was as genuine and unassailable; he would not have dreamed of laughing; she was as sensitive as he to the nuances of ridicule.

He lay down on the couch after removing his jacket and folding it over the arm for a headrest, and listened to the clicking of Miss Gold's claws against the tiles as she made her way back to her vigil by the door. It seemed to Chris to go on and on, the clicking, as if without him his darling found halls without boundaries to wander along; as if, without his presence in it, with his sense of fitness and order, the hall itself reverted to the native state of its materials and grew and grew, pushing out walls, lifting ceilings, until no barriers remained between what was once *here*, this safe place, and *there*, where all the unknown elements were.

He sat up, his fantasy having carried him into disquiet. His discovery that it was rain pattering on the terrace that had given rise to the fantasy of her journeying did not allow him to relax. He was first compelled, all logic to the contrary, to go to the door and peer down the hall to reassure himself that Miss Gold had not somehow contrived with her tricky intelligence to let herself into the public hallway where she could—and he was despairingly convinced, would, if she ever got out—be stolen by the first passing delivery boy. Her loss would be so mortal a wound—his imagination knew about the gradations in mortality—that all else beside it, all possible maimings, came to resemble remembered desire.

For the moment Miss Gold rested her tender burden on the cool tiles, tilted upward as though pausing midway in one of a series of pushups prescribed for overweight dachshunds. Chris's heart slowed with love. He blew a grateful kiss, and Miss Gold's left ear twitched as if the kiss had landed upon it.

Chris went back to the couch and lay down facing the windows, as though hopeful of finding answers or clues written on the sky. When he lay down the sky was all he could see. He could imagine into being whatever he wished below his windows, as he had done as a child. A trim garden had been his prospect then, intended by its seasonal blooming to instill in him an awareness of natural order, and into this garden he had summoned a nightly procession of disordered exotica. Below his window there had been Pyrenees and deserts, Petra, the Dead Sea. The muezzin had called there, jackals had fought, and ancient carp had swum in his garden.

Today he imagined the surge of the ocean, and the traffic obligingly metamorphosed itself into the distant roar of tides agitated by the pull of storm clouds. He felt safe, but felt it to be temporary. Anxiety hovered without, in the clouds, over the ocean, like a bird that would soon come tapping its beak upon his windows. But why did it wait for him, snug in his eyrie? Did it imagine that he would actually come and fling the casements wide, inviting it in?

His mind spoke to the bird: City apartments such as this one, owned pieces of the sky, are islands, separated from the mainland by an elaborate system of security. Even the wing-borne are eventually buffeted away, or if they hover too long, will find their pinions weighted by the sky-dweller's weapon, the formidable incinerator, and will plummet to earth to crawl dirty and miserable with the rest of the earth-bound.

The bird told him: Despite elaborate security systems and public weapons, sky-islands are frequently connected to the mainland by bridges of scandal and crime, over which the dirty, miserable, earth-bound pour in a stream, gawking, touching, destroying what is left by their envy and malice, and it serves the islanders right for inviting outsiders to cross over.

As I have done, you mean, Chris said. Scandal, crime—are these things usually preceded by premonitory dreams, such as the one I dreamed last night involving Victor Ramos?

In the dream Victor moved within a shadowy layer of odor. It was as though his shadow had been projected onto a surface some distance behind him, the shadow six inches fatter all around than the boy. This shadow was cut out and somehow imbued with the odors of oil, hot spices, armpits, rut, until it was malleable, something like silly putty, then it was reshaped around Victor, made somehow invisible and impermeable to water, soap, rainfall. In the dream Chris had seen it as a plain definition of Victor's unchangeable difference and his attraction for Chris and others like him.

But 'definition' meant nothing. Only the ingredients of the difference were known but not their proportions, nor how they were administered, nor why the knowledge was vitally essential to Chris. He knew with certainty that his life depended upon obtaining the formula. As though his desperation forced the issue, in immense close-up he saw huge needles like those with which the livers, kidneys, and hearts of embryonic sheep are imparted to decaying human bodies in Vevey. He saw phalanxes of white-swathed doctors and nurses at work making Victor's shadow pungent nearly beyond endurance. Chris cried out at the deepening of the mystery, cried with rage, demanded something in words that had no meanings. "Rut!" barked the doctor. Nurse and intern fell to a frenzy of fornication on the floor at the doctor's feet until Chris ground them to white powder beneath the heel of his shoe, handing, the meanwhile, the hypodermic to the doctor, for Victor's shadow had begun to neutralize in the moments of neglect, and without the acridity Chris would never have known the boy, and he could not bear to think of that. It was then, as though Victor read his mind, that the cruel Noh mask replaced the boy's features. Looming over Chris, the mask produced words from behind itself, clearly enunciated and yet not understood, though Chris had seen them written out as they were spoken: *¿Quere que te lo ponga en el roto?* Brutal, vulgar beyond comprehension, preceding the rest—the violence—

But Chris had tried too hard to look at the dream direct, and like a star it faded and was gone under his concentration.

Coldly, to counteract the thoughts that beaded his forehead with sweat, Chris took himself back to the night he had first met—or picked up; euphemism now could serve no purpose—the young Puerto Rican.

Unaccustomedly, Chris had walked from the house where he had dined to his apartment house, a distance of fifteen blocks. Fear of violence on the New York streets had long since settled into the subconscious, where it controlled routine. Once, in balmy weather, it had been necessary to invent reasons for not walking, but now the reverse was true, and Chris told the doorman, who was automatically whistling for a cab, that he was heading for a bar not two blocks distant. The man put up an argument to which Chris was, in his vulnerable state, a grateful listener until he saw that it was not concern for his safety but loss of a tip that brought the opposition, and the partial return to normality gave him the impetus to wave the cab away and cut across the street, leaving the doorman to dislike him as much as he wanted, which was no change at all in the night's status quo.

The recklessness that let him walk on the park side of Fifth, along the wall and past several of the park's gaping black mouths, was due to his having been for an entire evening the setup for a game whose denouement he had not allowed. He knew that no speculation now would ever unriddle it, and so he was frustrated as well as reckless, and fear, too, was pushing to the surface and eventually would have to be looked at.

His host had been a man to whom Chris had been presented some years before at the opera by a reprobate cousin, but a willing forgetfulness, based upon what he had felt to be mutual antipathy, had all but erased the meeting from time's record of occasions. And yet when he reecived the summons to dine—the man was famous and old and issued summonses—Chris recognized the voice before it identified itself, and he had gotten the clear impression that his own voice had been recognized by its 'hello.'

He had assumed that he would find his cousin, whom he hardly ever saw, in the company that filled the drawing room, if for no other reason than that Chris's telephone was unlisted, and from whom else could his host have gotten the number? But his cousin was not there, and when Chris mentioned his expectation—though, he implied politely, certainly not his disappointment—he had received a blank stare from the ancient yellowing eyes which denied the cousin's acquaintanceship and managed to reprove Chris for the gaucherie of forgetting through whom they actually had met. As if to make sure by paradox that the evening would be memorable for Chris through the early onset of total confusion, his host had introduced him into a group as the especial friend of someone of whom Chris had never heard. From his bluff height he had bent upon Chris a look of conspicuous dislike and had left instantly, a departure so abrupt that its rudeness could be called breathtaking. Finding himself in the double jeopardy of false pretenses and his host's challenge to him to reveal himself if he dared, Chris's disorientation was compounded by the strange effect upon the group of his introduction as 'a friend of ......'s.' It was as though the statement were both repellent and anticipated. The sound they made collectively was like the soughing of a copse in which the fox has hidden before the rush of hunters.

No effort was made to resume their conversation, and Chris, whose opening wedge, "Do you know my cousin?" had been taken from him along with his identity, gulped martinis from passing trays and played his part as well as he could. It seemed to him, upon whom alcohol had the instant effect of heightening prescience, that their eyes melted together, preparatory to forming a weapon of the resultant blue-grey metal, and he waited for the attack on whomever he was thought to be. When he did not run—for he imagined he could not—they reacted with a display of teamwork which bespoke an old alliance and probable masses of blue ribbons affixed to black velvet in shadow boxes on bathroom walls. Veiled, hostile remarks flew his way, to which, finding his footing, he reacted characteristically; then the veiling became latent and the hostility manifest. It was as though the point were to fix his outline, bristling with arrows, in space, to be used for their private purposes at some later time.

During the remarkably bad dinner, which he made no pretense of eating beyond the initial forkful of each course, it came to him that the gathering was archetypal, for a certain stratum of New York, in its composition, which was: cleverness, money, malice, estrangement, a sexuality so unsensual as to have the sharpness of weapons or acid. For him, the scapegoat that such gatherings demanded—or, barring the fortuitous stranger, created out of their own matter—there was deliberately fostered confusion which had been fed like an already obese animal all evening. In his deepening drunkenness he relived in lightning glimpses other such occasions to which he had been witness, or at which he had been the baiter or the victim. Tonight he played the game as he always did, as he had learned or perhaps been born to do, he did not care which. He insulted the food by refusing to eat it and degraded the wine by drinking it in large quantity without comment, as earlier he had inspected the pictures and *objets* in the drawing room with—not amusement, which would have been naïve—an artless, open indulgence. It was intrinsical to the players of such games that the illness produced by the adrenal glands in overactivity should be concealed; the stance of *cool* was a strict necessary discipline to hold in the vomit of fear and aloneness until one *was* alone. In the clarity of his alcoholic vision he saw that such parties were the roots from which the urge to violence seeped upward and out to undermine the city. Any literary gathering, he knew from experience, was the equivalent of a bomb factory.

He also knew that the denouement, whatever it was—which the players at that juncture might not have agreed upon—would occur when the coffee cooled untouched beside the sticky liqueurs, when the token splash of soda in the Scotch was abandoned like a final veil. When that time came he would know it by a fraction of a second before it was set in motion. In such timing lay the art of survival.

Even while the *frisson* of recognition of the moment coursed his skin, he was bidding his host good night, bending swiftly and up and gone toward the door before the old man could respond or try to hold him back. Leaving the room, he watched in a mirror both his approach and that of the hastily assigned huntress who was propelled from a group by a frantic push that caused her to stumble. She called after him, her voice strident, "We've just discovered who you are!" He admired the tactic, for ordinarily who would resist it? but he felt the acuteness of the danger when even a pause would be his undoing. Without stopping he turned his head only enough to project his words toward the silence, "Oh? But I don't know who *you* are," and made it without honor out of the room and into the elevator.

Once on the street and out of sight of the doorman's contempt, he gave way to the thought that had been pushing up: his knowledge of death had been somehow increased, as if a thinning of the wall of Death's room had occurred, against which he had been pressing his ear and listening to unimaginable preparations for a visit.

It was at that point of failure of his imagination that he saw Victor Ramos, standing with one leg encircling a fire hydrant, light from a distant streetlamp picking his bones clean. He was ostentatiously waiting for a bus; a sign around his neck could not have proclaimed it more clearly, especially to cruising police cars, than did his stance which, despite its leg-looped casualness, wore transience in every line as does a bird's on a telephone wire, with awareness of the hazards of high tension.

The image that reached Chris's brain was multiple, a straticulate cross section of an emotion, spheroidal by virtue of its two poles, ultimately static (it was Chris who circled the emotion, its restless satellite) because both poles were named 'delivery boy': a delivery boy; a dark-gold young man of ideal beauty; a skull; a statue of phallic symbolism (the hydrant between the legs) both lewd and de-sexed by antiquity, like a temple carving; a person hunted, without explanation, as Chris had just been; the embodiment of the type through whom Chris could restore his superiority and toward whom he could, therefore, be benevolent; a delivery boy.

Miss Gold was spending the night with Chris's housekeeper, the practice when he went out for the evening. Fear slept.

Chris caught the boy's eye, unsmiling, the game too serious for that. Then, apparently forgetting the exchange, he paused and took a cigarette case from the pocket of his dinner jacket, extracted a cigarette, tapped it on the case—(recalling, he saw the movement as calculated to draw attention to the case, which was platinum with a jeweled monogram and a concealed lighter which did not work)—and futilely flicked the lighter with the exact degree of impatience to seem to remain entirely private: no drop of self-amusement, with which the knowingly observed indicate the basic good humor of their natures.

Backbent over the lighter, Chris watched with a resurgence of foreboding a shadow advance on the pavement and cover his shoes and mount his ankles like dark secret water. When the boy's lighter snapped under his nose he jumped without pretense at the flame leaping in the dark hand. He drew on the cigarette, loathing its staleness (he carried the case for others, in all senses, and had last filled it in the fall) and expelled a cloud of flat smoke before he again looked at the other's face, finding shadow in which two hazy luminosities seemed no more committal nor unbenevolent than a moon divided into twins by capriciously unfocused binoculars.

It was soon clear that the boy, by lighting Chris's cigarette, had exhausted his part in the pickup, or the part he was willing to play. It was equally plain, because he did not move away, or speak, that he knew there was more to come, and the quality of his waiting bespoke experience at such encounters, which depressed Chris in some undefined way.

Reliving the night, Chris's depression took the form of Then and Now, for two reasons: the unresolved and the all-too-plain. He had always kept a special reserve of dislike for those relentless trackers of the virgin experience. Seeing himself among them, he could not find the courage of hypocrisy to make the distinction between their purely physical intent and his own. Chilled by hindsight, the most he could do for himself was to press on, knowing that the door to that room was now permanently ajar and that, sooner or later, he would have to go back to it, and enter, and take inventory.

Chris moved, the boy following, until the light was on the other's face. Watching him closely, Chris said, "*Yo estoy muy agradecido*," and felt ease settle over him at the expected reaction: the quick reappraisal of his clothes, skin color, and hair (the latter a repeated mystery until Victor explained about *good* and *bad* hair) by eyes in which a certain deference had been replaced by a like amount of familiarity dimly tinged with tentative contempt. To exorcise the contempt while retaining enough of the familiarity, which was a kind of trust, to make the rest possible: this was the delicate task, differing from case to case, which Chris felt lifted the moment above the potentially sordid.

It had been his belief, until the past few hours had placed it in escrow, that in that moment of assessment and decision by and through sensibilities heightened to the supranormal, he came to know more about the 'subject' as pertained to degree of trustworthiness and similar basic concerns than at any time that followed. That it was *his* decision, arrived at without help, had preoccupied him at the time: he had wondered how closely, under the circumstances, decision resembled invention, and therefore how much of what the subject subsequently said and did, apart from the seldom varying facts of life in the ghetto, was seen through the frame of that decision-invention, which he called the d.i. He imagined that the d.i. was like a picture frame, blocking out the nonconforming bits around the edges which otherwise spoiled the composition. But this line of questioning frequently led to the posterior thesis that the subjects *knowingly* conformed, an act of sophisticated duplicity which Chris could not concede them, as it gave them the clear advantage of anticipation and even led to the further question of who had invented whom. The thought that unriddling the maze of mirror images could bring him to a place where he functioned as the reflecting device—empty, unless a socially deprived, inferior person chose to stand before him—was enough to make him abandon the concept of the d.i., but his former preoccupation with it assumed negative coloration and hid out in his mind, another link in the chain of submerged fear.

He told the boy, "*Yo he bevido un poco*," met the now hostile eyes and said, concealing his distaste for the corrupted language which he had had to learn, "*Yo estoy fuquiao*." Hostility was replaced by astounded merriment at his admission to being fucked up, for the phrase, in street argot, was infinite, open-ended. It expanded his first admission to being a little drunk into a universe of implication, hinting, on a dark street past midnight, of the knowledge that drunkenness is but a station on the way and not the destination. The words were a clear invitation to camaraderie. Observing the facets of the statement, Chris rested.

The boy divined easily that the next step was up to him. He spoke with enough exaggerated humor to be able to point to the joke in case Chris should turn out to be a cop. 'Wha' you wan' me to do—walk you home?"

The surprises of the night prevailed. Chris did not know how to answer the question that cut so quickly to the bone of the matter. He had not thought beyond the game in any immediate sense and supposed that it was due to his real drunkenness that he had not. Staring blankly at him, he supposed that he had had in mind giving the boy his address and telephone number after some sort of exchange of credentials, a practice that was only a token, as their 'credentials' generally would not bear investigation. Also, the boy's answering him in English was like having a curtain behind which one was partially concealed pulled suddenly aside. He had been denied the cover of the argot before he was ready to relinquish it, and was being commanded to step forth as himself. Momentum had been lost through hesitation, and the boy was turning away to the curb. Chris sighed, hearing the released air brakes of an approaching bus as a gigantic echo of his signal of defeat—defeat, because the decision had been made for the first time by another person.

Chris said, a bit briskly for someone in his avowed condition, "Why not? We can have a drink, if you like, and I'll send you home in a taxi."

The boy raised his eyebrows, and it amused Chris to imagine that he could hear the boy's mind emphasizing the pride-offending word: "SEND?" He turned back to the bus, which was pulling in to the stop, but the quality of the movement was so unaggressive, even coquettish, that Chris knew one more word from him would be all that was needed. 'Please' or *'por favor'* would not do; simply saying *'amigo,'* which had worked for him once in a difficult situation, would be worse in the present one, for any note of pleading would revive the contempt, Chris felt, and rightly so.

Things were so far out of hand, as it was, that winning the game now consisted merely and entirely of keeping the boy from boarding the bus. After that, as far as Chris was concerned, he could go hang.

But the word? The boy gave it to him. The bus opened its doors, the few people inside gazed at the unlikely pair in the street without curiosity. The boy put one foot on the step, turned, and asked loudly, "You still wan' me to come wit' you?"

Chris flushed at how it appeared to driver and passengers and paid the price without smiling. "Yes."

Remembering, he saw again the amusement returning to Victor's face and with it a measure of admiration. Something in the look had made him feel his earlier drunkenness, about which he had forgotten through the minutes—emotionally, hours—so that when he turned to lead the way he staggered and Victor rushed to take his elbow and steady him.

They walked in silence for a block. Opposite his apartment house Chris said "There" and led the way across the street. The doorman on duty was the old man to whom his activities were never less than suspect. He thought, 'Let it all come down,' and smiled at Victor, who was taking in the building, expressionless.

Victor said, "Where you learn Espanish so good? You got Espanish wive?"

"No."

The doorman was holding the door open with mocking deference, his eyes on Victor.

"Girl fren'?"

"Nope. No boy friend, either." He ushered Victor in ahead of him, saying to the doorman, "This is—" and waited, head bent to Victor, who said with élan, "*Mister* Ramos," and the two of them proceeded to the elevator, united in a certain triumph.

And there it was, or the worst part of it, with no clue in the reliving to help Chris in his present quandary.

The rest of that evening had been predictable. Chris had once again been in control. He had explained over drinks about his 'job,' giving the title of his novel, *Minority Report*, as though it were a needed sociological document. He had watched Victor's adjustment to the fact that this was an encounter different from others that Chris was sure he had experienced. Chris had been told about such pickups and their aftermaths by similar young men who looked upon sex with other males purely—in an odd way the word was quite accurate—as a source of income. That night Chris had thought he could detect disappointment in Victor's behavior, smooth as the readjustment had been, that seemed to go beyond the idea of the money he might have been given, or stolen. Today, not entirely because of the reflected light of Victor's subsequently revealed antipathy to homosexuality that was like, in a complex way, Chris's own, Chris half believed, or half remembered, that what he had witnessed had been disillusionment. The idea fluttered in his mind, tantalizingly like a clue, but there was no place for it to alight. He hurried on in his mind, looking for some basis for the really odd thought.

From the time of their arrival in the apartment until the revelation that what Chris wanted of him was his hard-luck story, Victor had continuously assessed the apartment and its contents as if he were a prospective tenant. Following Chris's explanation, he had been silent for a while and then asked, "Why me?" He had laughed, unamused, and answered himself, "Because I was *there*, huh." He seemed willing to settle for it but Chris did not allow him to.

"Why did you come?"

"Man, everybody got a story. I never heard no rich man's story before. Me, I go along wit' *live*. ¿*Vida, si?* You wan' *my* story—" He shrugged. "May' I wan' *you* story. May' I writin' a book." He had seemed to laugh out of one side of his defiance, and Chris had laughed with him, relieved that Victor had not repeated his question of 'Why me?' Chris could not have answered it, and could not do so now.

Victor's beauty—certainly. He had never lied to himself, at least, about that. And yet, having had it to gaze at for long afternoons, he could not find wherein his satisfaction of it should lie. In frustration and what he feared might be honesty, he let himself imagine that it lay somewhere in destruction, in the destruction of the source, perhaps of beauty's source. Victor's death could not accomplish it; he would be equally as beautiful in immobility, the cold unmarred marble of his skin no more remote, no less beautiful, than when warmed by blood. And beyond that, beyond the marble's ultimate destruction by time, and beneath the marble and below time, the hard clean white bones, which were the broad preliminary but strong strokes over which the marvelous coloration and texture of his most overt beauty had been laid, hidden guarantees of the beauty's indestructibility. Even broken and crushed with a mallet, the bones still would sing in their dust and fragments, like pulverized jewels.

Chris's forehead burned his hand. It took effort to lift his arm, as if in reality that arm had wielded the mallet that crushed Victor's bones to an everlasting white powder. Everlasting. The source of beauty, then, lay in the opposite direction from the only one in which he could travel, in the country of death's antithesis, where he had never been and could not go.

"Nonsense." He sat up and mopped his hot dry face with a handkerchief as if his insanity had congealed on his face like a beauty mask. A corner of the handkerchief slid beneath an eyelid causing the eye to burn and stream water. Obscurely he said, "Beauty is in the eye of the beholder," and then understood his use of the cliché. He examined the handkerchief as though to find Victor there, safely, at last, removed from his body.

Leaving Victor and Miss Gold in the kitchen, on his way back to the study he reviewed his review, feeling that this time, unlike the mysterious baiting at the party the night he met Victor, there must be a denouement. He must, if necessary, force it, for Victor not only held the position that Chris had had assigned to him by the host that evening, but had held it—and *scapegoat*, he saw, was the only word—since their meeting. Chris had paid him to play the damned role and, fascinated, had kept him on long past his right to demand tenure. He saw that, infinitely more than the others, Victor had been meant to give him some key to himself. By his every action he had repeatedly asked the boy both 'Who am I?' and 'What am I?' There was subservience in the thought, and it rankled even more than he had imagined earlier, when he had let the supposition touch his mind: that until someone like Victor stood before him, he was an empty frame. The image he had used then was of a picture, but it was of a mirror he was forced, now, to think, for in Victor, somewhere, was the quicksilver that could throw back to Chris's eyes an accurate self-image.

At the study door he paused; Victor was not following him. Of the minutiae of sounds from behind the closed door of the kitchen he composed an acceptable picture of explanation: Victor, who really had been hungry but too proud after his outburst to go on and eat, was now eating the crucial loaf; apologetically, he was feeding small bits of meat and crust to Miss Gold.

Stepping into the study, it was as though he had tarried for an unmeasurable length of time in the hallway, for the sun was now out and glittered on the reservoir. A boat moved slowly along the submerged causeway that connected the pumping stations, sending up flights of gulls or ducks, he had never learned which, just as he had never learned the function of the boats. 'How little I know about my own surroundings,' he thought with cold distaste. Even the facts given him by Victor over many a long afternoon had been distorted by his imagination.

He stood at the window and watched the boat, saw that on each side of the turbulence of its wake there floated large circles of flat water to the edges of which the wake extended and nudged but did not break through. The reflected sun in the exact center of the largest gave it the aspect of one bespectacled eye, but it occurred to him that it and all the other rounds of water—he determined that they were not oil by minute inspection through the binoculars—were like the circle of which once he had been a part: self-contained, oblivious to other circles and the agitated, uncommitted swirling about it; and that the reflected sun in the center was the nucleus that each circle, of water or people, had to have to be autonomous. He thought it was like the finger that was needed to keep a ring from rolling away. For the first time without revulsion, he wondered if he and Victor and Miss Gold comprised a sort of circle.

As he watched, a weakness in the wall of the largest, sun-centered circle of water allowed a sharp little wave to skip through, and then another and yet another, each individually observed by him as they made their way to the sun-eye and sat upon it. Revulsion came then, because the effect was that of an eye covered with gold scales, and though the intruders did not retain their alien sharpness, but leveled out into glassy anonymity, he could not forget that they had been intruders, and he imagined that they affected the sun-eye's perception as though they were a rheum laid there by time and other excesses. The eye was lulled by its weight of gold beauty into an unwatchfulness through which a needle could pierce; perception would return a moment after blindness.

He stood at his bookshelves holding da Vinci's *Notebooks*.

THE LEAP OF WATER IS HIGHER IN A BUCKET THAN IN A GREAT LAKE. This is because (confined) water when struck by a blow cannot make its impetus pass from circle to circle as it would in a great lake; and since the water when struck finds near to itself the edges of the bucket, which are harder and more resisting than the other water, it cannot expand itself, and consequently it comes about that the whole of its impetus is turned upwards; and therefore water struck by a stone throws its drops up higher when its waves are confined than when they have a wide space.

"Yes," he said, and turned back a few pages and found what he had come to find:

No part of the watery element will raise itself or make itself more distant from the common centre except by violence.

He raised his eyes from the book and saw Victor, Miss Gold in his arms, quietly regarding him from the doorway. His face was troubled. Miss Gold struggled to get down with such determination that Chris wondered how Victor had offended her. The doorway framing them was like a page retaining the afterimage of what had just been read; Chris read the final sentence on the air: *No violence is lasting.*

Without a trace of friendliness he said, "Can't you see she wants down? Put her down."

Victor set Miss Gold carefully on the chaise, apologizing in a gravely polite voice, though the words, in Spanish, were too softly spoken for Chris to hear. When Victor picked up his glass, Chris turned away deliberately. After a pause Victor asked, "Drink?"

"No thank you."

"I mean for me." Chris let the silence deepen, staring at the reservoir, the circles, seeing the little high-ceilinged study as the bucket da Vinci had written about. He turned to Victor suddenly, as though to catch him at some misbehavior.

"I imagine you know where everything is by now." The subtle emphases, carefully placed, were as daring as he had ever been with someone like Victor, for they had almost the intimacy of equality. He was exhilarated and depressed. Victor's eyes slitted for a moment as though to improve vision.

"Not everting."

"Thing," Chris said, *"Ev-er-y-thing."*

Victor said, "Everting," flat as a slap, moved to the sofa, and sat in a series of gestures meant to be, and taken as, insulting. Ordinarily he edged his way in between sofa and coffee table, a space to which only the thin-legged should aspire; Chris had said that it was his housekeeper's way of trying to dictate the size if not the gender of his guests—but now the table was impatiently brushed aside with foot and calf and an aisle cleared by Victor. Then he turned and assumed a squat over the sofa. To emphasize his meaning he first looked back over his shoulder, then ducked his head and peered between his legs; finding something on the seat not to his liking, he straightened up, turned, brushed with vigorous scorn at the material, then turned again and fell back, legs thrown widely apart.

Chris snatched up Victor's glass, feeling fury at the pantomime and needing to get out of the room. Victor told him levelly, "Leave it," and when Chris stared at him, repeated, "Leave it." Chris set the glass down quietly and waited for directions, which Victor gave him in the same level tone.

"Sit down." Chris moved to skirt the table and sit on the sofa; Victor told him, "There. Opposite." Chris sat down in the opposite chair in which he sometimes read in the soft north light of morning, terrace door at his back.

Victor's movements continued in deliberateness which said 'Here is the end to pretenses.' When he had thrown himself on the sofa he had flung his arms up and back, clasping his hands over his head, resting them against the unprotected surface of a painting. Chris had thought of the hands sticky with food on the white-lit street of the Chirico; it had been a corner of his fury. Victor opened his hands, palms backward as though to smear the painting, then brought them slowly down to rest, equal partners, precisely sharing, on the perfect hemisphere that swelled at his crotch. His curved palms rode the swell, tightened only enough to reveal how thoroughly his manhood filled his big hands.

"Now," he said, and Chris felt his heart jolted by fear. There was a vacancy between him and Victor that was like the space between quotations marks set in the air; he expected the space to be momentarily filled by the words of the dream that he had not understood and yet understood: *¿Quere que te lo ponga en el roto?* Frantically he felt certain, he had to be certain, that if Victor *looked* at him, really *looked* at him, he would be able to read in his eyes without error his mistake about Chris. There had been what could be called flirtation between them, there had been sexual nuances, but in a sense that Chris had always known was Lawrentian: that male antagonism has a sexual base, and the more submerged, the more sexual, but—

Victor's sigh, loud and disgusted, saved him from what for all he knew might have been some overtness. With relief he saw that Victor's hands were moving, moving away, and then lay at his sides. As though demonstrating the wide possibilities of drama within the 'Now,' he said it again, "Now." He proceeded onward without the frightening pauses. "You ever had a fren' like me, Chris?" His enunciation retreated before his earnestness, consonants dropping by the way. 'Like' was 'li',' or as Chris saw it, inevitably, 'lie.' "You ever had a fren li' me? I mean, no money exchain hans—liddle loan between frens, but no pay for talkin'?"

"No." It was as though he agreed aloud with Victor, saying, 'Here is the end to pretenses.' He hoped Victor would not say, "Well, *could* you," for the answer—what was the answer? Miss Gold snored loudly. Chris gestured toward her—"There—" but Victor did not pursue the question by asking what he meant. Chris's 'no' had contained whatever he looked for—vindication? Instead, with softness more brutal than a shout he asked, "What do you *know*, Chris? You all the time correctin' *me*. Tell me what you *know*. What you learn in all these years about livin.' You not such a young man, Chris. You live a long time, you ou' to know plenny. Say what you learn that make goin' on livin' you right, man."

Chris felt dismay, but sensed that it was not only the inherent threat in the words that was the cause. He gazed at Victor, looking for the scintillae of reflected light to identify him. Victor was in no hurry; indeed, now that he had asked for something his face was suddenly anxious, like a good teacher's, that Chris should acquit himself well. Chris felt a dreaminess like an echo of the response to some other sound, someone else's need for him to bring off a situation well if not nobly. He thought, 'Hell, what *do* I know?' and spoke not as an answer but in answer to an inner voice.

"—*Je vois ma femme en esprit. Son regard, Comme le tien, aimable bête, Profond et froid, coupe et fend comme un dard; Et, des pieds jusque à la tete, Un air subtil, un dangereux parfum, Nagent autour de son corps brun.*"

"Beautiful, Chris. You wri' tha', man?" Victor's enthusiasm promised amnesty. Chris shook his head. Victor slammed his palm hard onto the sofa. "Dammit, Chris, tha' wha' the *man* learn from live, not you, Chris. A parrot, a stupido bird! can do what you do, Chris!" He shook his head, making a sound of disgust, "Chee!" He looked around as though for something to do with his hands, an exaggerated search, perhaps, but its franticness was plainly not all show. He grabbed his glass from the table and started up, violence in all his movements, in his restraint. At the door he turned and gave Chris a head-lowered long glance, and said in an orgy of discarded consonants, "I go' to thin' abou' this, man," made the sound "Chee!" again, and left.

For a comedian acquaintance Chris had once written a series of ideas of things people do immediately they think they are alone and unobserved, or, turning away in a crowded room, safely facing an unmirrored wall, the facial expressions they might register to relieve tensions. He was reminded so forcefully of these by his own actions when Victor was out of sight that he went about the grim business shaking with giggles. He moved the heavy cigarette lighter to the ledge of the bookcase nearest his chair, took the immense ash tray, a real murder weapon, and placed it on the window sill where a lunge might send him in time. Quietly, with a stealth he felt to be awful, he shifted the chaise so gently, moving it out of Victor's reach from the couch, that Miss Gold hardly noticed. When he had done these things he could not have told under oath exactly what had made him do them. Listening for Victor's return he sat down and reflected upon his predicament. If Victor came back and said, "Two things you've learned, Chris, in exchange for your life—" what could he offer?

He was smiling, rueful in self-knowledge, over the two answers that came out of a lifetime of experience when Victor returned: "Paranoiacs at bullfights identify with the bulls." "Left-handed people cannot tie a decent bowtie."

Victor came into the room wearily like a man moving under duress. He leaned over the table to set his drink down, and Chris saw the weight inside his shirt, narrowness above thinness was the way he mentally described it, and though he arrived quickly at what the weight was, when Victor pulled the butcher knife from his shirt the suddenness of the reality brought instant release. A torrent of sounds sluiced through the room. In shocked amazement, the heavy lighter in his hand, eyes fixed on the target between Victor's eyes, Chris first heard and then felt that the sounds were made by him. There was so much movement in the room, the scuttlings of dreams; and odors; and so much for the eye to encompass; and there was the sound jumping in his throat like a fish and the taste as it passed his tongue; and there was the gathering weight in his hand and the translation of it into arm movements. Victor came at him slowly, but his face remained his own—there was no Noh mask at which to aim. Chris thought two distinct thoughts simultaneously: 'I have never been so fully occupied' and 'I'll have to destroy his beauty.' With no further thought he pulled the lighter back and brought it forward with such force that, hanging onto it, he followed it to Victor. Victor threw the knife into a corner where it clattered on the floor, then Chris felt Victor's arms around him, keeping him from falling, following the trajectile of the lighter, onto the coffee table.

"Now," Victor was saying. "Now now now now." And "O.K., Chris? Come on, Chris." And "¿*Amigos, si?*" And "Now I can be you fren!." He murmured away, soothing and patting, "Fren, *amigos, si?* O.K.?" until Chris managed to shove him away. Without looking at him again, Chris found two tens and pushed them out toward the man, his eyes rising no higher than the beltless loops on the khaki trousers.

After a while the money was taken from him. He watched the hands fold the bills carefully and aim them toward a pocket before he turned away toward the reservoir. Thinking 'Oh, get out,' he waited out the long pause before he was obeyed.

He stood by the bookshelves, wondering about anger, about innocence, about essence. All of it was here to be studied, here in books, without ever going out again into the world. What wealth of real experience was here between book covers, tasted and tested by others, by others who knew how to translate and condense and present experience, knew how to account for their lives.

His hand touched a volume of Shakespeare's histories. He said, smiling for no reason he could have named, "Brief summers lightly have forward springs." Heavily, he said, "Likely, in this case; in his case," but it was not really appropriate. It was Victor who had acted Richard, and Chris who had been the young, inexperienced, noble York. Not; not noble; nor young; nor even inexperienced. Dangerously stupid—stupidly dangerous? He had come within an inch of smashing Victor's skull with the lighter. He could feel the desire and disappointment in his right hand, twitching on, as snakes are said to do after death.

Victor would get his, too, and soon. "Brief summers—" *Likely* was more hopeful than lightly. Yes, he would get his: an overdose, a knife in the belly—he would get his and soon. The vulgarism bothered him not at all; it was as though he were thinking in argot. Having suffered the ultimate indignity, the vulgarism nonpareil of seeing his own death before him, he could hardly be squeamish about slang, for all his old hatred of it, and thought of Victor getting his with comfort cold as ether on his skin.

He made himself stop it. He plumped pillows, emptied ash trays. Did the dead move so? Did dead minds twitch on remembering poetry? *"Viens, ma belle Chienne—"* was the way he always said it to Miss Gold, *"Profond et chaud—"* His bowels melted with the heat of horror. Where was she. A statue, he stood with the silent butler in one hand, Victor's glass in the other. In his stone head a memory tried to stir like a rock striving to create from itself a living worm: her claws clicking on the tiles beside Victor when he went to the door. And out. Both of them.

He crashed to the table the vessel that held Victor's breath imprisoned in alcohol as if it were the boy's body. The violence released his voice as the sight of the knife had done. Keening, he started for the door, discovered, by barely avoiding falling, that the only way he could walk was to place one knee-bent leg, foot pigeon-toed, laboriously before the other like a spastic.

He almost stepped on Miss Gold as she emerged trembling from under the chaise. His hands reached for her, two points of life at the tips of his dangling arms. She bared her fangs and backed away, under the chaise, a rumbling in her throat. Chris's spastic legs gave way, and he fell to the floor. The furious pain at the base of his spine was simultaneous with the intelligence that Miss Gold had moved her bowels under the chaise; he could smell and see the loose results, and this broke another obstruction in him. He crawled toward her, mumbling until the words he wanted formed themselves.

"You troublesome little bitch." Her ears lifted and folded back, and as his hand darted for her she snapped and caught in her teeth the fleshy pad that formed the base of his thumb, the part of the hand known as 'the mound of Venus.' The sick fear in her eyes and the sharp stab of her teeth revived, like summer rain, the parched part of his brain where pity grew.

"Don't be afraid, darling. Please. I won't hurt you." Soothing, cajoling, reassuring, he kept repeating the words until she was nestled in his nap and her trembling was subsiding. "Now, now, now," he told her.

He caught a glimpse, then, of himself in the mirror he had set up, and called Victor, and tried to shatter. Though he did not like it, hated the recognition and the implications of it, he paid it what respect he could. He told Miss Gold, making the mimicry as exact as his talents would allow, "Now I can be you' fren'."

MARIE, MARIE, HOLD ON TIGHT

HARRIET ZINNES

Here it is ten o'clock and the man who delivers the eggs has not appeared. He is usually so sallow that it is probably best he has the eggs in that white box. Even white eggs can look rusty yellow when a man in a long dirty white coat with a sagging face delivers them. Why I accept his deliveries I don't know. I don't even eat the eggs. I don't even cook the eggs. I don't even throw them away. I suppose Marie takes them. What other recompense does she have? I give her no money. She never asks for any. And is it right for me to be so crude as to suggest the need of lucre? Money. It is dirty yellow too. Marie likes bright colors. Every red or green dress I had she has taken. I have only blacks and whites left. She is selective. And she ought to be—with clothes, that is. She doesn't care about her men. Even the plumber, that fat, crooked-nosed, partly crippled man gives her titillations. I hear her giggle when he decides to take the rod, that long snakelike thing, to repair a stopped-up sink. She manages to get close to him, too. Asking silly questions. As if she cares whether or not her combings have stopped the sink. And he listens with that sly wink. Who knows what she does when she says good-by. She takes her time about it. When he comes I never seem to get my tea. And the toast is always burned that day. Maybe with the egg man too. Maybe that's why he's late. Maybe she's with him now. "Marie! Marie!"

Here it is ten o'clock and the man who delivers the eggs has not appeared. And I'm wearing Madame's low-cut red dress too, the one with the big bow at the place he likes so much. Such big hands he has. So much warmer than the plumber's, and quick too. He knows how to make me jump. It's good he comes in the morning. When I can get to that perfume and douse myself with it so when he takes off that dirty white coat and places his pants (neatly too!) on the chair (lucky he hates underwear) he can hold me tight and do it to me standing up so that I think I'll faint with pleasure— and we have to be so quiet too. Oh but what a nuisance is Madame, with her "Marie! Marie!" Does she know anything? What did that husband of hers ever do to *her* before she became an invalid. I wish I could have seen him. I'd give him his pleasures! I bet *his* was as big as that egg man's. At least his picture says all kinds of swishy bed things. I'd make him discover *me*. But I'm not doing too bad here: eggs all over me, and services of all kinds delivered here and to *me*. "Marie! Marie!" There she goes again.

Here it is ten o'clock and I'm still in my truck and I won't deliver the eggs. What eggs? I haven't any more to deliver. You'd think I gave Marie enough. Does she use eggs too? Insatiable that girl is. Insatiable. She tires me out. She thinks I'm a machine. An egg machine! Maybe I'll get her a pigeon. Pigeons make eggs and eggs make pigeons and pigeons make love too. And what about Madame. How long will she remain dumb? Or doesn't she care? I've heard stories. She knows too much. Her eyes can give a man the what for. She just sits in her chair doing nothing. Doesn't look at TV or read a book or knit. She doesn't even gossip. But those eyes of hers, giving me . . . Is it a come on? Yeah, her husband was a colonel. Those army wives. Maybe it is the come on she's giving me. With her sallow face. Her long dirty white dress. And Marie says she doesn't even eat eggs. Poison, she says. Poison. Maybe I can meet her on her own terms. **NO EGGS**.

IN CRYING'S BODY

MIA GARCIA-CAMARILLO

one. IN THE ABDOMEN.

Snail scream that you can't reach the sea
your one foot that is so slow you tie your hands around it to
pull on, foot you bang your head, slow, slow,
the sea going on without you—Mother! Wait!

In that little door she was. A house made of soot and the hundred windows, iron filings, mica, fool's gold, and inside was an iron mine for a cellar, twenty chimneys for a dining room, and all the laundry chutes the little one sleeps in (standing up), you, the dwarf, baby I had inside me for three weeks, I lost, that I beg you to tell me how your voice would have sounded . . .

In the white nightgown. Little she with oven-colored eyes, what are you standing in the door for! Fly to something! Go back in the chimney, fall somewhere, cry, look for me, make your fingers tick like a clock, try.

Don't stand there!

Don't stand there in me.

two. IN THE EYE.

a)   The pregnant skeleton's teeth chatter. She lived in a tin room. Down the corridor a green wool striped rug tried to feed her by her feet the warmth she couldn't get by electric bulb.

I want to make you a tin sun.

You know the doctor said to you, "O Maria! Why did you shoot up after you knew about the baby?"

When I held the baby for you, his eyes would shine like orange glass fruits. He knew. It would lie there on his collarbone like a spider or a suffocation.

Maria I don't know what you wanted me to tell him. That you were sorry because of sockets where your eyes had been? I was angry because you didn't care for him with the bones you had left. But even without flesh you still hurt and wanted a drug. You didn't ask for your flesh back. Maybe my anger was wrong. If your flesh had become him, you had a right to ask me this: to make him a picture, to call out his soul. "Make him a painting" you said. It was to go over his crib. I was too worried over what you would feed him to want to do this. I felt it was too much for me, but finally I was sad and wanted to.

I think she meant, "Make him a painting for a father."

b)   I painted him a dwarfed bird. I hung it over the baby's crib, but he wasn't there any more. Maria had become so drugged she had put the baby to sleep in the radiator. That's what she said. Throughout his childhood you could hear him crying in the radiator.

"Now I ask you God" I said "to make me very small. Because there has been no other way to get him out." But my request was not granted until a time people told me they had seen him, and I had no reason to doubt this.

three.   IN THE LUNG.

In my love's sleep I wandered, looking for him. I found his breathing, and I fell into his lung.

The first floor was a tiled garden with a skylight . . . The skylight in the rooming house in Paris where one laughs with joy because he knows he will be able to write there. Onion soup in the drugstore, the anarchist hotel clerk, tired doves in the clocks, and how people sense their thighs when they walk: and these mysteries without sounds are why we know that our suffering has not been for nothing, that a skylight is for flying out. So we can even sit down on tiles wet with fish tears and powdered scorpions.

If we fall asleep there or fall down it doesn't matter! We all think those things, and now I understand how a breath of air thinks them at the bottom of your cold, moody left lung. I can not speak about other lungs, but in yours I found blue stucco walls, windows grated with green iron, and in the w.c., the beak of a heron that once had got in, and whose flapping and scratching had caused you to mistrust your breathing for years. He is dead now, and it is his beak you feel, that you're afraid is a live tooth. But my love, I saw an alphabet he scrawled in pink blood when he got too weak to fly. For that memory I could say that a lung is not its skylight. Its floor is its mouth, desolate, permanent, with old songs that we need now, songs I am trying, that I have to keep trying to remember. Is to heal a scar to make it give up its song?

four. IN THE WOMB.

Is it something different, that we had a life coming to us, is it outside what went wrong like a skin is on the outside? My baby died when I had only been pregnant three weeks. It's not that I saw her, but just when they say I should have been scraped, a tiny umbilicus came out. What I want you to know is that the umbilicus goes with the small things here. I would like for you to look at your own capillaries, so that you can see this. I would like to remind you that my love and I are promised to each other. Your love is promised to you. These, and the pain of what is small to us, and of what in us is small to God, goes through our veins; and in women, floods the womb.

five. IN THE HEART.

When Maria's son was eighteen years old, his testicles felt the weight of his unused spirit. He had never loved anything, because his mother was a skeleton. His spirit tried to spiral out through his mouth in a kind of wild singing, and people were moved by this. But Maria's drugs were in his bones. The spiral fell back on him, weighting him stonier and more anguished between the legs, until something caught there and tore. From then on nothing went to life in Maria's boy. His hair fell out as if burned.

When I heard him scream I was afraid he could stay alive like that, screaming for years. The echo of his screaming was of a boy whose mother gave him away to a radiator. I wanted to be little again, like I could get him out, but by now he had fallen into a fire of so many parts of himself: his mother's fingernails, her hair, his own loose blown-off hair. So many parts of himself now that he had cancer smell for a cocoon.

I imagined he shrank as he died. I thought that he needed a tiny room in the underground, just large enough to emit this last light in, like he was a wick burning in oil. Let his ship of death be a lamp, since he and his mother had cold bones in life. But at this time it was out of my hands. And I didn't hear when his screaming stopped.

I knew it was good when he was gone. But in missing the place where my baby wanted to live I missed his not growing up. I can't know what wanting has to do with dying, or whether there has been some battle between the two. It's still hard not to think what took the dead ones away from us was bad, but wind isn't bad for taking leaves of a tree. What made my baby a person I cry for was this. She was the one who was tiny enough to get him out. Down below us, she flew a bird to the dark seed cave where he finally lived. I know this, because my little one was even more confused about these things than me; she didn't know where to put him after she got him out. Scared to have been the one that did what we couldn't do up here, my child, still smaller than an almond, *dropped* him. She had nowhere to put him except in me.

My breast aches. On this side it's started bleeding. Two sharp points I feel, you think they could be his hands piercing? Could he come out like shrapnel? Or is he growing and fattening? When I can't sleep I have thought I'll be another tomb for him to wake up in.

Things that go into us may come out again, but one thing I know is that I didn't tell Maria I couldn't help her. So I called that boy to me, and some of his end to me. I had a vein connected to the outside, to Maria too, I think it was because we're made this way. I am not safe, even in my heart. Lives have veins to each other.

Still, it's not Maria and her boy I want to grow in me, but my man. My baby will forgive me, that her death hurt me so I couldn't stand her dead. I am calling her, and now she must come alive. And she must take this disease and put it in the underworld. This time I will tell her where, it must have come from something like a deep pool, where it will blacken and fall apart. Then my baby will come back to me.

I am calling, and until she hears, it's bitter to me here. My blood loses its way, pushed out of its home.

Because my baby is not the one I am feeding, the one who eats: the ones who have been using my heart for burned food.

IRRITATIONAL VERSE

New Functional Poetry from Germany

Selected and translated by André Lefevere

Most introductions are notoriously expendable. This one is no exception. The poets grouped together in this anthology are perfectly able to speak in their own voice. All the translator/anthologizer might profitably do for the benefit of a non-German-speaking audience is to try to fill in a bit of the background against which they are writing. The poets presented here belong to a brand of German poetry which has been consistently underestimated by the critics, academic and otherwise, who are ultimately responsible for the canonization of what usually gets to be called "the main literary current of an age." These poets take their bearings from a tradition of "functional poetry": poetry, that is, for everyday use, poetry on everyday subjects, poetry which is, above all, critical of the everyday life regulated for poet and reader alike by very real social and economic—as opposed to more shadowy "mystic" or "supernatural"—powers over which they have little control.

Functional poetry goes back, via Brecht, Tucholsky, Kästner, Fried (to list only its main representatives in more recent years) to the naturalist and socialist poets of the nineteenth century, and beyond them to the often anonymous popular literature which has provided an alternative to the more consciously literary, highbrow, aestheticizing tradition ever since the Middle Ages. Twentieth-century functional poetry is neither the "cosmic," dislocated poetry of the Expressionists, nor the hermetic, esoteric poetry associated with Rilke. The poets whose work is translated here are very much concerned with Germany's (on both sides of the border between East and West) here and now. Their verse is critical of both the "new Germanies" all of them grew up with. After the tremendous economic upswing and the recent anarchic eruptions of deep-seated discontent it produced, today's functional poets sit back for stock-taking. The result is a low-brow, terse, elliptic, epigrammatic poetry that has little to do with the "belles lettres" so often used, if not designed, to maintain the status quo. It is a poetry written not for its own sake, but to put things, as Karsunke has it, "into context," to make the reader aware of the social and economic context in which he lives and, more particularly, of the shortcomings to be found in that context; to foster, in other words, and on a large scale, the critical awareness that is a prerequisite for any lasting change not imposed from above. It is, consequently, a poetry which changes the reader's part in the reading process from a more passive "experience" to an active dialogue, a thinking along with the author on very concrete matters of common concern.

Yet these poems are no versified one-page recipes for a revolution "devoutly to be wish'd"—there are traces of resignation and despair in some of them, of fear of being absorbed by the publishing industry, of being academicized or rendered inoffensive in other ways; some ask questions about two of the constant factors in man's existence that escape all social codification: love and death, and quite a few are concerned with the possible function of literature itself.

The poets grouped together here are the latest exponents of the "functional school" in modern German poetry. Brecht, Tucholsky, Ringelnatz are history; Fried and Kästner belong to a previous generation; Biermann, Degenhardt, and Süverkrüpp are more successful in a different medium, that of the song. The poets in this anthology all started publishing in or after 1965, and they have not, so far as I know, found their way into English much beyond the occasional poems translated in magazines like Dimension or MPT—and yet their work could, I believe, become "functional" also in the development of contemporary verse written in English, because it shows the way to a poetry that can be critical of the society in which it originates without degenerating into cheap sentiment, empty sloganeering, narrow-minded propaganda, or hopelessly frustrated vituperation.

A. L.

TO THE GERMAN READERS
AND THOSE WHO WRITE FOR THEM

PETER SCHNEIDER

Let me try to describe the appearances for a moment. I don't see washing on the lines when I look out of my window in the morning, no children on the balconies, no walls with Castro, Mao, Ho Chi Minh painted on them, ten times over, with red paint. I see the caretaker's little garden; I don't know him, I only know him from his little notes: keep the stairs clean, keep your doors locked, take off your overcoats, floor just polished.

I see the caretaker's little garden, cut up into perfect rectangles and kept so clean that nothing grows in it. I see the wall, two meters high, around the little garden, and twelve meters long, and on top of the wall I see broken glass, to keep the neighbor's children out. When I look across the courtyard, away from the wall, I see another, bigger wall, also armed with broken glass, to keep the neighbor's children out. But when I look all over the courtyard I can't see the people the walls were built against. I don't see children storming walls, no son sets a car on fire, no fathers get machine guns and collapse on the asphalt with their sons' arrows in their bodies. I see cars on the asphalt, cars belonging to people who don't know each other, people I don't know, people who only know their cars and only get to know each other because of their cars. I see windows, and I know the people who live behind them only because they pull their curtains when they see me at my window—and they know me only because I pull my curtains when I see them at their windows. I see these windows open, not infrequently, at night, and they are filled with the faces of people who have worked for eight hours and relax by making sure that nothing will happen. And all this doesn't just happen to me, but to all of us.

When I go out into the street I don't see people mixing with each other, no groups of people talking about the newspaper, no conversation at all. I see people who look as if they were living underground, as if the last time they enjoyed themselves was their child's third or fourth birthday. They move as if surrounded by a network of electric wires that sends shocks through them whenever they dare raise an arm or swing a foot. They walk past each other and look at each other as if each were the other's mortal enemy. The impression you get of life here is that there must be a big war on, somewhere, and that everybody is waiting for the all-clear signal to move again.

When I walk into a bakery I am very careful not to put my hands on the glass counter, because it has been pointed out to me that the counter might collapse. When I point at a cake I am very careful not to stretch my hand out too far, because it has been pointed out to me that I might infect it. When I pay, I take great care to put the money on the little rubber sheet, because it has been pointed out to me that that is why the little rubber sheet is there. And all this doesn't just happen to me, but to all of us.

When I happen to be waiting somewhere, together with somebody else, we don't look at each other, we don't touch each other, we don't enter into any relationship whatsoever. I once sat in a full waiting room for three hours, among people with the same background and the same worries, and not a word was spoken, until somebody opened a door marked: "No Entry"—everybody jumped up and shouted: "No Entry." People's way of life here is so bad that they show their fists to express a need for affection: all they mean when they shout at you is that somebody should take some notice of them at last; their wishes and their interests have been frustrated and neglected so often that they take anybody who crosses the street when the lights are red for a would-be assassin.

When I look for an example to describe what happens in our traffic, I am reminded of a scene I once saw in a film made in the thirties: Hardy bumps into Laurel's car, Hardy walks over to Laurel's car and pulls off a fender, Laurel smashes Hardy's steering wheel, Hardy tears Laurel's engine apart, Laurel slashes Hardy's car seats, Hardy rips the top off Laurel's car, Laurel sets Hardy's tires on fire, Hardy pushes what is left of Laurel's car into the next wall, Laurel finally cuts down a tree and the whole thing ends with a hell of a bang.

I am reminded of that scene, not because it doesn't correspond to anything in reality, but because it does: I quote the following account from a newspaper. Mr. X, from Berlin, almost hits Mr. Y, from Wuppertal. Both gentlemen stop their cars at once. Mr. X walks over to the enemy's car and hits Mr. Y on the mouth, through the open window. Mr. Y pursues the fleeing Mr. X and plunges his pocket-knife in X's lung. Mr. X, seriously injured, grabs his wife's umbrella and starts chasing Mr. Y around the car. Mr. Y calls on his wife for help. Mr. X hits Mrs. Y over the head with his umbrella. Mr. Y comes to the rescue. Mr. Y, Mrs. Y, and Mr. X fight with pocket-knife and umbrella until all three of them collapse, dead.

No, I can't see anything on these streets, on these balconies, in these houses, behind these windows, that looks vaguely human. These houses have not been built for people; they are inhabited by strange wallpaper, paintings, furniture, and TV sets that keep telling their owners that they own the right wallpaper, the right furniture etc. Streets are not there for people to tell each other what they want and don't want: they are dominated by cars that take their drivers to places where they are exploited on week days, and are repaired and polished by them on Sundays. Neon lights don't say what the people want, but what Osram, Siemens, and AEG would like to see happening.

If I could see something else, then I would describe something else. It's not my fault if I don't compare these streets to a jungle; I've got enough imagination to detect motherly feelings hidden in a prostitute and if, seen from the ninetieth floor, this town reminded me of some soul's skeleton, I can assure you I'd notice.

I might stuff myself with marijuana and hijack a plane to Cuba, if I didn't know I've got to do the same thing, but with insight and out of hatred. I'd put a real gold and silver crown on the Shah's head, because I think you'd hardly notice him otherwise, if I hadn't convinced myself that he should be eliminated. I might compare capitalism to a world-famous cyclist, whose tendons have been so twisted by his exclusive devotion to his sport that he can neither walk nor sit down, if only capitalism hadn't twisted my own tendons to such an extent that I can only fight against it, not compare it to anything whatsoever.

I would bum around cemeteries in the middle of the night, dig up Heinrich von Kleist's poor skeleton and read him Brecht's poems, to make him laugh a bit. I would do all this and more I have forgotten because it's too unrealistic.

When I look around, my eyes don't seem to find anything that would give me joy—all I can see is the struggle of all against all, and so I have to fight too, I am dragged down all the time, never lifted up, not a single one of my senses and desires is emancipated, they are repressed all the time, to end up where they were when I was five years old.

I have only told you about very simple, very obvious things, and

I have given you a few examples only, because everybody can put two and two together. Somebody else could start from another spot in this society, and he would reach this, or a similar conclusion.

It is, of course, possible to record the history of a broken marriage, if the writer still believes that his story is the least bit different. Or else somebody could write his biography if he is still inclined to give the individual a bigger part in literature than he gets in society. We can, indeed, go on for another ten years, each and every one of us, go to pieces on the job, in our marriage, in our families; we can go on being surprised to see the strain of work increase, because work is becoming more and more superfluous; we can go on being surprised to see that the nearer our wishes come, objectively speaking, to fulfillment, the more we are left, alone with our wishes only, and we can, each and every one of us, go on describing this misery and reading about it for another ten years, as if we were faced, time and again, with a fate so immovable that only our attitudes toward it can change. And yet we should have noticed that a few boxers and rockers we know are the only ones who have been certified to exhibit a certain non-neurotic tenderness—because they hit back, hard enough, every day.

No, what we see around us, what we experience, can no longer be described; it can only be changed. Every now and again we meet a genius, who gets away from it all and presses a few liberating images out of his recklessly isolated imagination. But who for and what with do they write, these writers, sculpt, these sculptors, compose, these composers. They show us a world in movement, but in reality the world gets more and more static every day; they appeal to the aesthetic sensibility of people whose primary needs are so frustrated that they will even put up with appeals to aesthetic sensibility. They write concerts for ears made deaf, day after day, by the authoritarian orders given by their superiors; they play on an imagination which allows itself to be possessed by lofty thoughts only—if all goes well—in the theater, and only to fall back on thoughts of possession straightway; they are a challenge to eyes which have never had a chance to look at the products of their work other than as "goods."

Then there are the realists, who have long ago come to terms with capital's subjugation of their creative faculties. They are capable of one reaction only against their misery: to admit it any time and to describe it as conscientiously as they can. They live in the silent hope that a description of his misery will rouse the reader to a fury they themselves prefer to invest in rousing descriptions. This literary nightmare is just one more addition to the many subtle nightmares everybody experiences and takes in their stride every day, on the job, on the streets, in the movies. What could be stirred in the reader has to be brought out into the open first—through action.

There are, finally, the avant-gardists who "solve" the conflict between human wishes and a hostile reality by not acknowledging it in the first place: they use all their energy to represent methods of representation, not unlike people who have gone mad waiting for their order and start eating their knife and fork with knife and fork, instead of the roast that fails to materialize—after which they leave the table with all the symptoms of immense satisfaction.

All these forms of art have, taken as a whole, lost their traditional function: to protect human wishes from capitalism—on the contrary, they protect capitalism against the rebellion of human wishes. These artistic expressions no longer express the promise of a future fulfillment of human wishes; they work the other way around: they change the real misery into a kind of promise, by turning it—on top of everything else—into an object of the imagination.

All this is by no means a question of talent, and no talent can protect itself against this evolution. But when imagination is so completely driven out of society that art becomes the representative of the bureaucracy in the realm of creation, then the time has come for wishes and desires to break their form as art and to look for a political form. Imagination can only survive if it really conquers the ground that was taken from it, not in the realms of imagination, but in reality.

Does this mean the death of literature? No, all talk about the death of literature has always been put into this world by those who are only looking for an excuse to present bourgeois literature in a new, unheard-of form of rigor mortis.

Does this mean that bourgeois literature is dead? Yes, bourgeois literature, which separates the despair of the great majority from its explicable and analyzable causes, and twists it into an "enjoyable experience," a literature which is able to express, without the faintest indication of surprise, nothing more than resignation, renunciation, and loss in the midst of abundance, a literature which shows the masses their misery only to make them grow accustomed to it, such a literature is dead and must be buried. Whenever somebody describes the oppression he suffers from, he should also describe the alternative, the causes of the oppression and the strategy that will lead to liberation.

When a total mutilation of all human wishes corresponds to the total development of methods of production in the last phase of capitalism, then art has to be the total mobilization of all wishes against reality. It should, therefore, do two things: portray the wishes and also portray capitalism, oppose concrete images of reality against concrete images of the potentialities hidden, suffocating in it. Human wishes must, in this confrontation, be kept as free as possible from their artistic form, so that they can find their political form. It is not the task of art to organize human wishes artistically but to draw them out of repression, to guide them toward revolution in their natural state.

Mao Tse-tung has described an example of this kind of art. When on the Long March, the Red Army used to organize meetings of farmers in the market places of the villages it passed through. One or more farmers were asked to describe their personal wishes and worries, and the oppression they suffered from. The farmers in the audience took part in this self-description, by criticizing it: they completed it when it seemed too personal, and corrected it when it seemed too general—in short, they turned it into a collective drama. They called it "The Great Lament," after which the Red Army offered itself as a form of struggle and organization for needs that were, at first, very basic and unpolitical. The form corresponding to these needs was therefore neither prayer nor confession, neither poem nor novel, but the Red Army. This model is, of course, applicable to our own situation, because our workers and farmers are not better, but worse off than their Chinese counterparts. They don't die of hunger, they die secretly of oppression and humiliation. Let us try the "Great Lament" in factories, schools, and universities. Let us cultivate the workers', students' and pupils' inability to suffer oppression, as well as their ability to smell it from afar. It is the task of the artist—if he happens to be a person who has not let capitalism corrupt his imagination completely—to help workers, students, and pupils articulate their wishes and to show them the way to political organization.

PROMETHEUS 70

Kurt Bartsch

Only
when the central heating
is off
or coal deliveries
far behind
does a slight displeasure
cloud the brow
of the man chained
to his own house.

POOR SOD

Friedrich Christian Delius

One night at two we stormed the critic's
well-known house. He was still at work,
jumped up with great relief
and raised his hands at once.
Watched, shammed a little gleeful indignation
as we packed his books in wicker baskets,
but did not give us a hand. We remembered
his well-known enthusiasm for "La Chinoise"
and left him with Brecht and Mayakovsky.
He brought some wine.
When we got to the records all he said
was he'd gone off Beethoven anyway but he suddenly
insisted on Albert Ayler. We took a vote
and left them behind. We danced with his wife
who invited us to the kitchen for snacks
we ate with perfect table manners.
He wanted to keep us with whisky after that
and when it was getting light and

we dragged the stuff outside
he asked us to come again.
That, we thought, went too far.
Where did we go wrong?

EXAM QUESTION

Yaak Karsunke

what do you do
when she gives you a gun
(i.e. the revolution)
the doorbell rings
there she is
asks for fire
no lighter & no matches
explosive speeches fiery exhortations
she wants
fire & offers you the gun
fire & you'll be free

even though you close the door
as quick as you can without a word
you know she'll still be standing there
(i.e. the revolution)
& it won't be easy
to slip past her on the steps
to get out into the freedom
of the unfree
you can't turn tail
you can't buy her off

when she gives you a gun
what do you do?

HOLLOW SHAPE

> but the cup of white gold at Patara Helen's breasts
> gave that—   —Ezra Pound

Yaak Karsunke

the white gold of her breasts
probably gave
shape to the cup at Patara
death to Troy &
a pretext to the Greeks

it gave blind Homer breath
for (at least)
fifteen cantos
and blinded Pound
an image for two lines

almost as beautiful
as Helen's breasts
—as the cup at Patara—
and almost as empty.

IRRITATIONAL VERSE

Friedrich Christian Delius

"I see," says Wolgang Maier in the local,
Uhlandstrasse, one night in July sixty-five,
yes, I see, that's the way to write a poem,
I see, start just like that and then
count all the things I see, starting with
the business section of the paper for all I care
there's the poetry of this world for you,
just quote and say I see, Delius, why don't you

write poems like that, long compositions,
says Maier, write something like that and say I see
and then say what you see, get cracking, take
the talking here for instance, take the subsidiary
clauses from the conversation there with Piwitt
and Buch and Born, Franz and Sarah,
get cracking, I see, I see, he says, I'll write you
a poem like that right now, I see the onion
on the tuna salad, the juke box frothlike on the waves,
why else do the Beatles sing in our ears,
Uhlandstrasse is what I write and newspapers trampled on,
rain too, a little touch of the lyrical,
schnapps I see even though there is no schnapps
just when I'm really in form he says,
had Doornkaat and milk for breakfast
was at the airport tonight anyway, Delius,
say I see and then say what you see,
pick it up in the trams and on the streets
in old pubs, in new pubs, anywhere
on the bus, with the whores, with the queers,
with Jack and with Jill, with liquor and beer,
at home or in the university or wherever you get it,
write, write it down, write us a big long poem
show us you're not short of breath, anything you see . . .
everybody's writing long poems these days
so don't try to get out of it,
Höllerer is not under discussion it's my idea anyway
three years ago on the plane we used to
talk a lot about long poems and Whitman
and Pound and all the vigorous bearded bards
who could still do it, but all you do
is stammer your eight piffling lines
the Germans have no poetry I said
because they're out of breath and Höllerer agreed
on the plane, we should write long poems he said
and long poems he wrote, the best people nowadays
all write long poems, who are they anyway
the idiots who write short lyrics, all lame in the loins
I tell you, Delius, throw away your matches and
take a deep breath and say I see and
I see and open your eyes take the world

into your poem he says, we need a new
Storm and Stress, that's right says Buch
from his corner, a load of crap says Born, let them talk
says Maier, just start writing, nothing pretentious
look here, I had a whole liter of milk for breakfast
today and a bottle of Doornkaat and two rolls,
that's nothing special, but I just wanted to tell you
and I was at the airport tonight
lying in the grass off the runway with a girlfriend,
an old one, and talking about old times
that was quite something, something to remember
says Maier, show us what you can do,
get cracking, take a deep breath, don't get stuck
on ideas, just start writing, show me what you can do
show me when I'm halfway sober, a long poem
is what I want from you, Delius, I'm telling you,
I tell you it's the only way to show how good you are
write, show me, I want to see it, see, I see, No
said I, Maier, not like that, not
like that

POSTSCRIPT

Say listen, says Wolfgang Maier three years later
when I was on the plane with Höllerer
some time ago—we were flying to Zürich—
I found myself thinking of your damn poem all the time
and I couldn't say a word, Delius, I just sat there
looking out of the window with nothing to say
just because of your stupid poem, then,
a stupid feeling
          Is that right, Maier, says I
and then I get cracking with my two sentences
on the function of poetry and
so on

ON CHARLES OLSON, DEAD

10.1.1970

YAAK KARSUNKE

in addition to his research on maya
hieroglyphs in yucatan: in addition
to his intensive research
on massachusetts' early history
: the professor wrote poems
crammed with cultural
history
history culture

the *main source of*
one of his poems
no kidding
is John Burnet's
Early Greek Philosophers
(london 18/92)
: there is a poem for you
from another:
"a son of the working class
illuminates
like a slot-machine"
the author himself remained obscure
"in his lifetime already
a myth
in american poetry"

this *damn intellectual twit*
who bought ready-made suits
(his own point of view)
died in a hospital in new york
he left a gap behind
& reading matter.

FLOWER POWER

YAAK KARSUNKE

Lenin had had it
in 1917
he decided on that famous
flowery nonviolent march
on the Winter Palace
(which has since gone down in history)
walked at the head of the column himself
crowned with corn-poppies, smiling
munching sunflower seeds of course
typically Russian

the czarist troops scattered
sheaves of sharp shooting
over the demonstrators mowing down
Lenin & Trotsky, Sinovyev,
Kamenev, Radek, Bucharin—all of them
shot dead—small consolation
that comrade J. Stalin, too,
went to his rest

: but beautiful to the present day
is the garden of zarskoje selo
where the crown prince Romanov
(not unlike the boy
who beheads thistles)
gets his early training in pulling out weeds
while his father, still smiling,
cuts away the roses,
the red ones, still.

HYMN

Friedrich Christian Delius

I'm afraid of you, Germany,
word invented for our fathers,
not for us
         you with your deadly hope
you in the coffin doubly blackened,
Germany what am I to do
with you, nothing, go away
leave me alone,
Germany you are stoning us again
running to your death over
double-edged tongues
double-edged swords
I'm afraid of you, Germany,
I beg of you, go
leave me the language and go
you, wide between the marks, rotten already
and still not dead, die, Germany
I beg of you, leave us alone
and go.

SLOT

Johannes Schenk

She's got a mattress from Mexico full of straw
doesn't get up before noon, looks into the letterbox
if somebody's in there or if there's a flower
in the slot between her legs
my many letters.
She's got breasts of shell-stone, they don't rush
like the others, no tritons' horns, they tick-tack
as if there's somebody inside, knocks or wants out
and two red light-buoys in the dark above
I'm a sailor, I know.

BERLIN

after water paintings by George Grosz

KURT BARTSCH

Green the streets, a purple girl
stands opened by the walls
gets plugged by an accountant
at short notice

Street lamps rattle, snow, music spreads itself
out of green windows, Jazz and Charleston
too late for the soldier—two legs amputated

The bridges scream, gas meters breathe
offer death to a suicide
back from the night shift in despair

The roofs, coffins, lie on the town
walls underneath, moon, a throat slit
and policemen on the beat

The tram fails to bridge the misery
runs through houses, night
tears through the curtains
caught behind all the applause.

SINS BABEL

Yaak Karsunke

> *Isaac Babel*
> *born 1894*
> *arrested 1938*
> *liquidated probably 1941*
> *rehabilitated 1954*

1

for Isaac
& others who are nowhere
now
not a finger was lifted
then
—or if it was
"requested to deposit the ashes
in the appropriate container"

poets too are mortal
many now know the sundial's sentence
the uncertain hour
of their certain death
of which there are more versions
than of their texts
(once more immortal)
reprinted
an example for many

their life too now exemplary
maybe their death
(facts are sadly lacking)
their heritage at least
is honestly divided
the scars for one side
the other keeps the knife.

2

Black I see black on white
the names of my friends
in a newspaper—
not as new by far
as its name pretends—
the names of friends
oppressed in print

doubt is cast on their doubts
their skepticism skeptically observed
in this other country praised too soon
(in which they use the same words
as we do here)
my friends stand
in the night behind reflecting windows
see mainly themselves
and through that everything

at eye level
they see Isaac's eyes
and in front of their lips
the shadows on the corners of his mouth
where laughter lies (buried
or hidden)
from there it will come
over the living and the dead to judge
lift up make known or what have you

JEWISH FRIENDS

Yaak Karsunke

i have no Jewish friends
my mother was Jewish
said one, the other said
i'm Mosaic
when asked about his religion

& that was their way
not to talk about it.
i don't care, i said,
race is a myth
& to hell with nations
that was my way

my parents agreed that
it was easier for us
all: just one world war
instead of two
& still children in
those terrible days with the Nazis
you felt a touch of melancholy
for your own childhood
when you listened to them (only then):

"Jackboot's gotta die-hai-hai
too young to cry-hai-hai
Jackboot's gotta die-hai-hai
too young to cry-hai-hai"
—the shoes of my childhood
are worn down
they fit a few things:

my father was an antisemite
but my mother, e.g.
sacrificed her Jewish friend
as she puts it
so's not to endanger us
my brothers & sisters & me
i.e. she stopped seeing him
when things got dangerous
& brought my two sisters unendangered
into this world
he was still alive then
until the others came
to get him & he did not come back

my mother still weeps for him
things are easier for me today

never faced with this alternative
I have no Jewish friends
just a skeleton
hung in my cupboard
years later
that's what my mother
did for me
& nowadays nobody could begin to imagine
what it meant then—as she puts it

: the shoes of my childhood
are worn down
a few things fall through—
"if the sole kne-hu-huw
Jackboot's in the to-hu-humb
Jackboot's gotta die-hai-hai
too young to cry-hai-hai"

: there must be tons of shoes left
in the former camps in the East
my mother won't go there
there is no need to—as she puts it
she's got enough imagination
which seems to be defective only
in the case of cypresses, e.g. & orange groves
so that's where she goes to
& sends me color post cards
to help me imagine, too

you buy slip-soles
when there's a hole in your shoe
to keep the dirt out a bit
things were easier for me
& my coast is clear
cleared by my parents
when I was still a child
I have no Jewish friends
one friend's mother was Jewish
the other often says Mosaic
I'm not buying slip-soles
I'm not buying myself out.

KILROY WAS HERE

Yaak Karsunke

when i was 11
"Kilroy was here"
was written on cracked walls
on fallen columns
tables in cafés, bogs,
the amis
wrote it everywhere

when i was 11
my sisters wore red skirts
my mother herself
had ripped off the white circle
with the cross broken
in four places
& burnt it
& now Kilroy was here

when i was 11 the war was
over & Hitler kaputt
like the houses, the windows, the Jews
& Germany (what was that?)
but Kilroy had come instead
taught us basketball
gave us chewing gum & coke

when i was 11 Kilroy
taught me words like fairness
& democracy
slogans like no more war
jitterbug
& even with Shakespeare sonnets
a Brooklyn accent

when i was 11 Kilroy is here
were three golden words
almost as beautiful as the three

of the French revolution
he talked about
liberty&equality&fraternity

when i was 11 my parents
had given me
the wrong education
Kilroy tried hard
explained human rights
& the un charter
to re-educate me

when i was 11 Kilroy
was my best friend
his house was open
& in his cellar
i heard jazz & Stravinsky
& no sirens

much of that was left behind
—years later—
when Kilroy got into his plane
stuffed it with napalm & disappeared
now Kilroy is here
is written on pagodas
& black smoking ruins

we
are separated.

MY GRANDMOTHER, E. G.

Kurt Bartsch

They didn't put my grandmother
e.g. in the museum
although she was a washerwoman

six-armed goddess
of the streets

There was no Alexander available
when she got all knotted up with gout
unlike in Gordium e.g., in the old days
so she went to the welfare doctor
who found grounds for hope

He prescribed a protracted cure
in Arcadia e.g. after which
my grandmother studied local transport routes
(thoroughly, I'm told) but failed to find
the promised land

So she had no other choice
than to put up with mountains of dirty underwear
and kick the final bucket
when the pain became unbearable
(Tantalus e.g. a slight ache
by comparison).

THE CHAIR

Kurt Bartsch

Death takes paper and a pencil
and draws a chair, empty
before empty windows, a chest

spilling over empty jackets
and a clock that asks the time.
Empty too the bed and skinned already.

Death takes paper and a pencil
draws a chair, empty
and sits for long hours before the fire.

GRAVE, DIG?

for Rainer Hachfeld

Yaak Karsunke

I don't know
how long it takes
a man to die
I don't know how
he falls
with a bullet in his belly
or in his head

all i know
are urns & coffins
literature & cinema
Lessing e.g. says
the Greeks represented death
as the brother of sleep

"crumpled up
in that clothes-bag position
which always means the same thing"
says Raymond Chandler.

BIRTHDAY

Friedrich Christian Delius

My fear has a birthday
in February
I drink with my enemies
all night long
to celebrate
and each in turn we interpret
pale oracles
from empty glasses.

The guardians of order, too
are my enemies.
They decree a beautiful
state of emergency
while we give a cheer
three cheers for my fear.

The next morning
my enemies scare my hangover away
(time they earned their invitation)
with congratulations
they hope I'm satisfied and
that I'll put it back now,
with all their gifts, my fear
into my heart.

THE LION OF JUDAH

for Martin Buber

JOHANNES SCHENK

The Lion of Judah feeds on straw in the circus
with people applauding all around
dust falls from heaven through the food-hole
but where is peace?

Is there not peace anyway
as long as there is no war?
Why the fuss?
they sit in peace.

Peace sits in the bar next door,
"The Crown," drinks aquavit
smashes windows, burps
and fights around with the police.

The lion in the cage
would like to walk through town
yellow as the sun at noon
to hold the Torah

—they call far over the sea
the lighthouses,
because there is so much black water
and not a single lamp—

for the rabbi
in his soft paws
he's given his claws to the children
for necklaces.

UNDERGROUND

JOHANNES SCHENK

The bakers have started preparations
to sell their stuff—bread, rolls, pretzels, cheesecake—
to the fish on the bottom of the sea
mine fumbles around with a snorkel in his garden.
The cobblers learn how to sole fins
in various sizes and the few dopeys,
nobles, kings, and town clerks, learn
to bridle torpedo-fish, the twits.
Only the women will be saved: no more washing ever.

The neighborwomen are already in training
to scheme under water as on earth
they crochet slips, underwear, and men's flies
with seaweed, ship's yarn and herringbone.
The three traditional parties, i.e.
the priest-, business-, and puppet-party
are looking for water-resistant paper
for their posters but the coal merchants
want briquettes that burn under the sea

for grandmothers and babies in their cribs.
Masons and locksmiths start building escalators
on the shore and dumps for traffic signs.

Lawyers have started quarreling over mixed marriages
(What with dolphins, mermaids, and sirens?)
Seahorses draw the local bus
and the dead are buried in the air above.
Anarchists and pacifists sit in their bars hoping
that the powers that be generals and policemen
will not go down with us, that they'll find it
too wet maybe and get scared of sharks
                                              their counterparts.

BIOGRAPHICAL NOTES

KURT BARTSCH. Born in Berlin, 1937. Studied at the Institute for Literature, Leipzig, 1964–65. Lives in East Berlin. Poetry published: *Zugluft* (1968), *Poesiealbum* (1968), *Die Lachmaschine* (1971).

FRIEDRICH CHRISTIAN DELIUS. Born in Rome, 1943. Has lived in West Berlin since 1963. Spent most of his youth in Hesse, one year in London. Poetry published: *Kerbholz* (1965), *Wenn Wir, bei Rot* (1969).

YAAK KARSUNKE. Born in Berlin, 1934. Various jobs. Editor-in-chief of *Kürbiskern*, 1965–68. Lives in West Berlin. Poetry published: *Kilroy & andere* (1967), *reden & ausreden* (1969).

JOHANNES SCHENK. Born in Berlin, 1941. Ex-sailor, lives in West Berlin. Poetry published: *Fisch aus Holz* (1967), *Bilanzen und Ziegenkäse* (1968), *Zwiebeln und Präsidenten* (1969).

PETER SCHNEIDER. Born in Lübeck, 1949. Has lived in West Berlin since 1961. Published: *Ansprachen* (1970).

SUMMER TIDINGS

JAMES PURDY

There was a children's party in progress on the sloping wide lawn facing the estate of Mr. Teyte and easily visible therefrom despite the high hedge. A dozen school-aged children, some barely out of the care and reach of their nursemaids, attended Mrs. Aveline's birthday party for her son Rupert. The banquet or party itself was held on the site of the croquet grounds, but the croquet set had only partially been taken down, and a few wickets were left standing, a mallet or two lay about, and a red and white wood ball rested in the nasturtium bed. Mr. Teyte's Jamaican gardener, bronzed as an idol, watched the children as he watered the millionaire's grass with a great shiny black hose. The peonies had just come into full bloom. Over the greensward where the banquet was in progress one smelled in addition to the sharp odor of the nasturtiums and the marigolds, the soft perfume of June roses; the trees have their finest green at this season, and small gilt brown toads were about in the earth. The Jamaican servant hardly took his eyes off the children. Their gold heads and white summer clothing rose above the June verdure in remarkable contrast, and the brightness of so many colors made his eyes smart and caused him to pause frequently from his watering. Edna Gruber, Mrs. Aveline's secretary and companion, had promised the Jamaican a piece of the "second" birthday cake when the banquet should be over, and told him the kind thought came from Mrs. Aveline herself. He had nodded when Edna told him of his coming treat, yet it was not the anticipation of the cake which made him so absent-minded and broody as it was the unaccustomed sight of so many young children all at once. Edna could see that the party had stirred something within his mind for he spoke even less than usual to her today as she tossed one remark after another across the boundary of the privet hedge separating the two large properties.

More absent-minded than ever, he went on hosing the peony bed until a slight flood filled the earth about the blooms and squashed onto his open sandals. He moved off then and began sprinkling with tempered nozzle the quince trees. Mr. Teyte, his employer and the owner of all the property which stretched far and wide before the eye with the exception of Mrs. Aveline's, had gone to a golf tournament today. Only the white maids were inside his big house, and in his absence they were sleeping most of the day, or if they were about would be indifferently spying the Jamaican's progress across the lawn, as he labored to water the already refreshed black earth and the grass as perfectly green and motionless as in a painted backdrop. Yes, his eyes, his mind were dreaming today despite the almost infernal noise of all those young throats, the guests of the birthday party. His long black lashes gave the impression of having been dampened incessantly either by the water from the hose or some long siege of tears.

Mr. Teyte, if not attentive or kind to him, was his benefactor, for somehow that word had come to be used by people who knew both the gardener and the employer from far back, and the word had come to be associated with Mr. Teyte by Galway himself, the Jamaican servant. But Mr. Teyte, if not unkind, was undemonstrative, and if not indifferent, paid low wages, and almost never spoke to him, issuing his commands, which were legion, through the kitchen and parlor maids. But once when the servant had caught pneumonia, Mr. Teyte had come unannounced to the hospital in the morning, ignoring the rules that no visits were to be allowed except in early evening, and though he had not spoken to Galway, he had stood by his bedside a few moments, gazing at the sick man as if he were inspecting one of his own ailing riding horses.

But Mrs. Aveline and Edna Gruber talked to Galway, were kind to him. Mrs. Aveline even "made" over him. She always spoke to him over the hedge every morning, and was not offended or surprised when he said almost nothing to her in exchange. She seemed to know something about him from his beginnings, at any rate she knew Jamaica, having visited there three or four times. And so the women—Edna and Mrs. Aveline—went on speaking to him over the years, inquiring of his health, and of his tasks with the yard, and so often bestowing on him delicacies from their liberal table, as one might give tidbits to a prized dog which wandered in also from the great estate.

The children's golden heads remained in his mind after they had all left the banquet table and gone into the interior of the house, and from thence their limousines had come and taken them to their own great houses. The blonde heads of hair continued to swim before his eyes like the remembered sight of fields of wild buttercups outside the great estate, stray flowers of which occasionally cropped up in his own immaculate greensward, each golden corolla as bright as the strong rays of the noon sun. And then the memory came of the glimpsed birthday cake with the yellow center. His mouth watered with painful anticipation, and his eyes again filled with tears.

The sun was setting as he turned off the hose, and wiped his fingers from the water and some rust stains, and a kind of slime which came out from the nozzle. He went into a little brick shed, and removed his shirt, wringing wet, and put on a dry one of faded pink cotton decorated with a six-petaled flower design. Ah, but the excitement of all those happy golden heads sitting at a banquet— it made one too jumpy for cake, and their voices still echoed in his ears a little like the cries of the swallows from the poplar trees.

Obedient, then, to her invitation, Galway, the Jamaican gardener, waited outside the buttery for a signal to come inside, and partake of the birthday treat. In musing, however, about the party and all the young children, the sounds of their gaiety, their enormous vitality, lung power, their great appetites, the happy other sounds of silverware and fine china being moved about, added to which had been the song of the birds now getting ready to settle down to the dark of their nests, a kind of memory, a heavy nostalgia had come over him, recollection deep and far-off weighted him down without warning like fever and profound sickness. He remembered his dead loved ones. . . How long he had stood on the back steps he could not say, until Edna suddenly laughing as she opened the door on him, with flushed face, spoke: "Why, Galway, you know you should not have stood on ceremony. . . Of all people, you are the last who is expected to hang back. . . Your cake is waiting for you. . ."

He entered and sat in his accustomed place where so many times past he was treated to dainties and rewards.

"You may wonder about the delay," Edna spoke more formally today to him than usual. "Galway, we have, I fear, bad news. . . A telegram has arrived. . . Mrs. Aveline is afraid to open it. . ."

Having said this much, Edna left the room, allowing the swinging door which separated the kitchen from the rest of the house to close behind her and then continue its swing backwards and forwards like the pendulum of a clock.

Galway turned his eyes to the huge white cake with the yellow center which she had expressly cut for him. The solid silver fork in his hand was about to come down on the thick heavily frosted slice resting sumptuously on hand-painted china. Just then he heard a terrible cry rushing through the many rooms of the house and coming, so it seemed, to stop directly at him and then cease and disappear into the air and the nothingness about him. His mouth became dry, and he looked about like one who expects unknown and immediate danger. The fork fell from his brown calloused muscular hand. The cry was now repeated if anything more loudly, then there was a cavernous silence, and after some moments, steady prolonged hopeless weeping. He knew it was Mrs. Aveline. The telegram must have brought bad news. He sat on looking at the untasted cake. The yellow of its center seemed to stare at him.

Edna now came through the swinging door, her eyes red, a pocket handkerchief held tightly in her right hand, her opal necklace slightly crooked. "It was Mrs. Aveline's mother, Galway. . . She is dead. . . And such a short time since Mrs. Aveline's husband died too, you know. . ."

Galway uttered some words of regret, sympathy, which Edna did not hear, for she was still listening to any sound which might try to reach her from beyond the swinging door.

At last turning round, she spoke: "Why, you haven't so much as touched your cake. . ." She looked at him almost accusingly.

"She has lost her own mother. . ." Galway said this after some struggle with his backwardness.

But Edna was studying the cake. "We can wrap it all up, the rest of it, Galway, and you can have it to sample at home, when you will have more appetite." She spoke comfortingly to him. She was weeping so hard now she shook all over.

"These things come out of the blue," she managed to speak at last in a neutral tone as though she was reading from some typewritten sheet of instructions. "There is no warning very often as in this case. The sky itself might as well have fallen on us. . ."

Edna had worked for Mrs. Aveline for many years. She always wore little tea aprons. She seemed to do nothing but go from the kitchen to the front parlor or drawing room, and then return almost immediately to where she had been in the first place. She had supervised the children's party today, ceaselessly walking around, and looking down on each young head, but one wondered exactly what she was accomplishing by so much movement. Still, without her, Mrs. Aveline might not have been able to run the big house, so people said. And it was also Edna Gruber who had told Mrs. Aveline first of Galway's indispensable and sterling dependability. And it was Galway Edna always insisted on summoning when nobody else could be found to do some difficult and often unpleasant and dirty task.

"So, Galway, I will have the whole 'second' cake sent over to you just as soon as I find the right box to put it in. . ."

He rose as Edna said this, not having eaten so much as a crumb. He said several words which hearing them come from his own mouth startled him as much as if each word spoken had appeared before him as letters in the air.

"I am sorry . . . and grieve for her grief. . . A mother's death. . . It is the hardest loss."

Then he heard the screen door closing behind him. The birds were still, and purple clouds rested in the west, with the evening star sailing above the darkest bank of clouds as yellow as the heads of any of the birthday children. He crossed himself.

Afterwards he stood for some time in Mr. Teyte's great green backyard, and admired the way his gardener's hands had kept the grass beautiful for the multimillionaire, and given it the endowment of both life and order. The wind stirred as the light failed, and flowers which opened at evening gave out their faint delicate first perfume, in which the four-o'clocks' fragrance was pronounced. On the ground near the umbrella tree something glistened. He stooped down. It was the sheepshears, which he employed in trimming the ragged grass about trees and bushes, great flower beds, and the hedge. Suddenly, stumbling in the growing twilight he cut his thumb terribly on the shears. He walked dragging one leg now as if it was his foot which he had slashed. The gush of blood somehow calmed him from his other sad thoughts. Before going inside Mr. Teyte's great house, he put the stained sheepshears away in the shed, and then walked quietly to the kitchen and sat down at the lengthy pine table which was his accustomed place here, got out some discarded linen napkins, and began making himself a bandage. Then he remembered he should have sterilized the wound. He looked about for some iodine, but there was none in the medicine cabinet. He washed the quivering flesh of the wound in thick yellow soap. Then he bandaged it and sat in still communion with his thoughts.

Night had come. Outside the katydids and crickets had begun an almost dizzying chorus of sound, and in the far distant darkness tree frogs and some bird with a single often repeated note gave the senses a kind of numbness.

Galway knew who would bring the cake—it would be the birthday boy himself. And the gardener would be expected to eat a piece of it while Rupert stood looking on. His mouth now went dry as sand. The bearer of the cake and messenger of Mrs. Aveline's goodness was coming up the path now, the stones of gravel rising and falling under his footsteps. Rupert liked to be near Galway whenever possible, and like his mother wanted to give the gardener gifts, sometimes coins, sometimes shirts, and now tonight food. He liked to touch Galway as he would perhaps a horse. Rupert stared sometimes at the Jamaican servant's brown thickly muscled arms with a look almost of acute disbelief.

Then came the step on the back porch, and the hesitant but loud knock.

Rupert Aveline, just today aged thirteen, stood with outstretched hands bearing the cake. The gardener accepted it immediately, his head slightly bowed, and immediately lifted it out of the cake box to expose it all entire except the one piece which Edna Gruber had cut in the house expressly for the Jamaican, and this piece rested in thick wax paper separated from the otherwise intact birthday cake. Galway fell heavily into his chair, his head still slightly bent over the offering. He felt with keen unease Rupert's own speechless wonder, the boy's eyes fixed on him rather than the cake, though in the considerable gloom of the kitchen the Jamaican servant had with his darkened complexion all but disappeared into the shadows, only his white shirt and linen trousers betokening a visible presence.

Galway lit the lamp, and immediately heard the cry of surprise and alarmed concern coming from the messenger, echoing in modulation and terror that of Mrs. Aveline as she had read the telegram.

"Oh, yes, my hand," Galway said softly, and he looked down

in unison with Rupert's horrified glimpse at his bandage—the blood having come through copiously to stain the linen covering almost completely crimson.

"Shouldn't it be shown to the doctor, Galway?" the boy inquired, and suddenly faint, he rested his hand on the servant's good arm for support. He had gone very white. Galway quickly rose and helped the boy to a chair. He hurried to the sink and fetched him a glass of cold water, but Rupert refused this, continuing to touch the gardener's arm.

"It is your grandmother's death, Rupert, which has made you upset. . ."

Rupert looked away out the window through which he could see his own house in the shimmery distance; a few lamps had been lighted over there, and the white exterior of his home looked like a ship in the shadows, seeming to move languidly in the summer night.

In order to have something to do and because he knew Rupert wished him to eat part of the cake, Galway removed now all the remaining carefully wrapped thick cloth about the birthday cake and allowed it to emerge yellow and white, frosted and regal. They did everything so well in Mrs. Aveline's house.

"You are . . . a kind . . . good boy," Galway began with the strange musical accent which never failed to delight Rupert's ear. "And now you are on your way to being a man," he finished.

Rupert's face clouded over at this last statement, but the music of the gardener's voice finally made him smile and nod, then his eyes narrowed as they rested on the bloodstained bandage.

"Edna said you had not tasted one single bite, Galway," the boy managed to speak after a struggle to keep his own voice steady and firm.

The gardener, as always, remained impassive, looking at the almost untouched great cake, the frosting in the shape of flowers and leaves and images of little men and words concerning love, a birthday, and the year 1902.

Galway rose hurriedly and got two plates.

"You must share a piece of your own birthday cake, Rupert . . . I must not eat alone."

The boy nodded energetically.

The Jamaican cut two pieces of cake, placed them on large heavy dinner plates, all he could find at the moment, and produced thick solid silver forks. But then as he handed the piece of cake to

Rupert, in the exertion of his extending his arm, drops of blood fell upon the pine table.

At that moment, without warning, the whole backyard was illuminated by unusual irregular flashing lights and red glares. Both Rupert and Galway rushed at the same moment to the window, and stared into the night. Their surprise was, if anything, augmented by what they now saw. A kind of torchlight parade was coming up the far greensward, in the midst of which procession was Mr. Teyte himself, a bullnecked short man of middle years. Surrounded by other men, his well-wishers, all gave out shouts of congratulation in drunken proclamation of the news that the owner of the estate had won the golf tournament. Suddenly his pals raised Mr. Teyte to their shoulders, and shouted in unison over the victory.

Listening to the cries growing in volume, in almost menacing nearness as they approached closer to the gardener and Rupert, who stood like persons besieged, the birthday boy cautiously put his hand in that of Galway.

Presently, however, they heard the procession moving off beyond their sequestered place, the torchlights dimmed and disappeared from outside the windows, as the celebrators marched toward the great front entrance of the mansion, a distance of almost a block away, and there, separated by thick masonry, they were lost to sound.

Almost at that same moment, as if at some signal from the disappearing procession itself, there was a deafening peal of thunder, followed by forks of cerise lightning flashes, and the air so still before it rushed and rose in furious elemental wind. Then they heard the angry whipping of the rain against the countless panes of glass.

"Come, come, Rupert," Galway admonished, "your mother will be sick with worry." He pulled from a hook an enormous mackintosh, and threw it about the boy. "Quick, now, Rupert, your birthday is over. . ."

They fled across the greensward where only a moment before the golf tournament victory procession with its torches had walked in dry clear summer weather. Galway who wore no covering was immediately soaked to the skin.

Edna was waiting at the door, as constant in attendance as if she were a caryatid now come briefly to life to receive the charge of the birthday boy from the gardener, and in quick movement of her hand like that of a magician she stripped from Rupert and surrendered back to Galway his mackintosh, and then closed the door against him and the storm.

The Jamaican waited afterwards for a time under a great elm tree, whose leaves and branches almost completely protected him from the full fury of the sudden violent thundershower, now abating.

From the mackintosh, however, he fancied there came the perfume of the boy's head of blonde hair, shampooed only a few hours earlier for his party. The odor came now swiftly in great waves to the gardener's dilating nostrils, an odor almost indistinguishable from the blossoms of honeysuckle. He held the mackintosh tightly in his hand for a moment, then drawing it closer to his mouth and pressing it hard against his nostrils, he kissed it once fervently as he imagined he saw once again the golden heads of the birthday party children assembled at the banquet table.

THREE POEMS

JOSE HIERRO

Translated by Louis M. Bourne

TRANSLATOR'S NOTE: *José Hierro's poetry provides eloquent moral testimony to the sadness of self-defeated Spain after its grim Civil War. "Irremediably, we postwar poets had to bear witness," he writes. Whether dealing intimately with his sense of lost ideals and opportunities (influenced by a four-year prison term for his Republican sympathies) or publicly with his historical context, he is unavoidably elegiac. His poems lament both the waste and the irrevocability of time.*

Born in Madrid in 1922, Hierro grew up in Santander and studied industrial engineering there until the Civil War. There too, between 1945 and 1952, he edited, with poets like José Luis Hidalgo, the literary journal Proel. *He presently lives in Madrid, where he works in a publishing house and teaches.*

Hierro's first two books appeared in 1947: Tierra sin nosotros *("Land without Us") and* Alegría *("Joy"; awarded the* Premio Adonais). *These were followed by* Con las piedras, con el viento *("With the Stones, with the Wind," 1950),* Quinta del 42 *("Call-up of '42," 1953), and his* Antología poética *("Collected Poems," 1953), which won the* Premio Nacional. *Three other books and another collection comprise his output so far:* Estatuas yacentes *("Reclining Statues," 1953),* Cuanto sé de mí *("As much as I Know of Myself," 1957; awarded the* Premio de la Crítica *and, in 1959, the* Premio March), Poesías completas *("Complete Poems," 1962), and the* Libro de las alucinaciones *("Book of Hallucinations," 1964).*

Hierro divides his poetry into two classes, "reports" and "hallucinations." The first are narrative; the second, mysterious in the way they call up emotions with only a vague reference to the events that inspire them. Despite his often rich poetic expression, Hierro considers his language to be "dry and bare" and feels his poems should be remembered not as literature but as moments in the reader's own life.

FOR AN AESTHETE

You who smell the flower of lovely words
Perhaps do not understand mine with no aroma.
You who search for water that runs crystal clear
Do not have to drink my red waters.

You who follow the flight of beauty perhaps
Never thought how death walks the rounds
Nor how life and death—water and fire—together
Are undermining our rock.

Perfection of life that carves and readies us
For the perfection of distant death.
And the rest, words, words and words,
Oh wonderful words!

You who drink wine from a silver cup
Do not know the course of the spring that bursts forth
In the rock. You do not satisfy your thirst in its pure water
With your two hands cupped.

You have forgotten everything because you know everything.
You see yourself master, not little brother, of all you name.
And you forget the roots ("My Work," you say); you forget
That life and death are your work.

You have not come to earth to place dams and order
In the wonderful disorder of things.
You have come to name them, to identify with them
Without raising fences to their glory.

Nothing belongs to you. Everything is flowing, a stream.
Its waters empty into your temporal riverbed.
And becomes a single river, you all pour out into the sea
"Which is death," the stanzas say.

You have not come to place order, a dam. You have come
To make the millstone grind with your transitory water.
Your aim is not in yourself ("My Work," you say); you forget
That life and death are your work.

And that the song you sing today will one day be silenced
By the music of other waves.

REPORT

From this prison you might
See the sea, follow the whirling
Of seagulls, feel
The pulse of living time.
This prison is like
A beach: everything's asleep
In it. The waves break
Near its feet. Summer,
Spring, winter,
Fall, are roads
Outside that others walk:
Things with no effect, changing
Symbols of time. (Here
Time is meaningless.)

This prison was first
A cemetery. I was a child
And sometimes came by
This place. Shadows,
Cypresses, broken marble.
But already rotten time
Contaminated the earth.
The grass then wasn't the cry

Of life. One morning,
With pickaxes and shovels
They disturbed the freshness
Of the ground, and everything—niches,
Rosebushes, cypresses, walls—
Lost their old pulse.
They raised a new cemetery
For the living.

From this prison you might
Touch the sea; but the sea,
The newborn hills,
The trees that die out
Among yellow chords,
The beaches that open huge
Fans to the dawn,
Are external things, things
Without effect, ancient myths,
Roads that others travel.
They are time,
And here it is meaningless.

Otherwise everything is
Terribly simple.
Water in the morning has
A fountain's shape . . .
     (Water-taps
At daybreak. Bare
Backs. Eyes wounded
By the cold dawn.) Everything
Is simple here,
Terribly simple.

And so the hours pass. And so
The years. And perhaps one warm
Evening in autumn
(They speak of Jesus) we feel
Time suspended. (Jesus
Spoke to men and said:
"Blessed are the
Poor in spirit.")

But Jesus is not here
(He left by the big stained-glass window,
He's running along a cliff,
He's going in a boat with Peter
Over the peaceful sea).
Jesus is not here. The eternal
Fades away, and it's the ephemeral—
A blonde woman, a misty
Day, a child stretched out
On the grass, a lark
That tears the sky—it's the ephemeral,
What happens and what changes,
That holds us captive.
A thirst for time, because time
Here is meaningless.

A man goes by. (His eyes
Full of time.) A living being.
He says: "Four, five years . . . ,"
As if he were throwing the years
Into oblivion.
A boy from the valleys
Of Liébana. A country person.
(He seems to hear the voice
Of his mother: "Son,
Don't be late," the dogs barking
Among the green pines,
The blue flowers of April
Budding . . .)
        He says: "Four, five,
Six years . . . ," serene, as though
He were throwing them into oblivion.

The sky, sometimes blue,
Grey, purple or glowing
With lights. Sometimes golden.
Scattered, heavenly gold.

We know very well who scatters
The gold and gives the lily
Its vestments, who lends

The wine its red color,
Who soars among clouds, ordains
The seasons . . .
                (Roads
Outside that others walk.)
Here time is without symbols,
Like stray water which the river
Never shapes.

And I walk among the things
Of time, I come and go on, lost.
But I am here, and here
Time has no meaning.
An angel, immortality lost,
With the nostalgia of a little grain
Of time. On seeing me, they think:
"Perhaps he's asleep . . ."

Because without a proof
Of time, I'm not alive.

From this prison you might
See the sea—I don't think now
About the sea—. I hear water-taps
At daybreak. I don't think
The running water sings to me
The fountain's cold song. I cut
My new paths.

So I won't feel alone
Forever and ever.

EXPERIENCE OF SHADOW AND MUSIC

Homage to Handel

It was not the divine music
Of the spheres. It was another
Of a human sort: of air, water
And fire. A music with no hours

And no memory. Flesh and blood
Without beginning or end. Vault
Of nocturnal larks. Honeycomb
Of flame on the faraway peaks.

I remember it perfectly.
Shining, through the grace and
Power of mystery. Transfigured
From eternity and fever and shadow.
It was an impossible music
Like a living creature. Marvelous
As a moment become eternal
In its zenith. I heard its burning
Waves. With my fingers, I felt
The quivering of its form.

Here time begins. Urn
Of moon, prison of fragrance.
Now everything is celestially
Real. Viola-veins play,
Horns—nostalgias, hearts—
Oboe-carnations. . . Who strips
The leaves from the subterranean light,
From the harmonious numbers? What chords
Rob life from the silence, melody
From the flesh, kisses from mouths?
Glass of centuries from the fountain
From which all silence springs.
You too, my daughter, music,
You too . . . ?
        Eagle, roving
Crown, you too? Magical,
Alone, majestic, high above,
Unmoving, do you reign, do you
Rule the night? . . . And you come down
To the rock where the Promethean flesh
Suffers its old nomadic thirsts.
And you sink your bill into its entrails,
You torture it till it begs.

Of earth and air and water
And fire and flesh and blood . . .
Marvelous as a moment forever
Present. Drop by drop, you
Drink the resonant stars;
Sip by sip, all the pain,
All the life, all the dreaming:
The Universe. No matter now
To die, to make ourselves
Your echo. Death cuts through sadness
With its prow; you are its wake:
Pulverized light. You sink deep
Into the soul: you make it more soul;
Into the frozen flesh: you restore
Its spring, you dress it with soul,
Linking it to your orbit.

It was not the celestial music
Of the spheres. A thing
Of our world, it was death
In motion. It was the shadow
Of death, paralyzing
Life at the edge of the dawn.

And suddenly the silence is heard.
Everything recovers its own light.
The flesh—our flesh heard—goes back
To being stone, prison, grave.
I sank my hands of diamond
Among the pale corollas.
I lifted up the crests of water
To the kingdom of the gulls.
Hands that had crossed many
Miles of waves. That had been,
For only an instant, a burning
Mouth against another mouth.
That had been life, and were
Clouds and ashes in the memory.

Fatal shred of beauty,
We can only cry alone.
But now without tears, without
Words, the mysterious words
That say what they hide,
That silence what they reveal.
Without clarity if seen.
Made of granite when touched.
Fatal shred of beauty
That vanishes when named.
Petal by petal, it withers
On the frost of music.
Feathery arcs snatch it away . . .

And the night once more regains
Its reality of pale ruins
Beneath the light of the torches.

EMBER DAYS

E. M. BEEKMAN

Mr. Van Tussen had never married. He had not been able to incorporate the possibility, not to mention the fact of marriage into his unadulterated existence. It would be a crime if he were to destroy the delicate clockwork of his present life. This remarkable achievement was partly due to the fact that Mr. Van Tussen had never worked in his life.

His father had been the owner of a successful company which manufactured precision instruments. The company had been small but renowned. His father had told him that during the annual meetings of The Society of Precision Instruments Manufacturers, his company had been hailed (remember boy, year after year) as the national answer to Swiss hegemony. This was, naturally, before the time his father sold out to a foreign firm. The father retired to live abroad until his death. But he had insisted that his only child would receive a proper education in his native country and had put him into the hands of his sister when the boy was four years old. The aunt, a spinster, was experienced in raising boys, having been the principal of an exclusive boarding school. The creased, tallowy lady had granted him a proper education in morality, in manners and in laconic conversation. Thanks to this solid foundation, Mr. Van Tussen (at that time a quiet and wan child) struggled admirably well through the nation's primary and secondary school systems, finishing with honors and ready for the educational icing to be provided by the capital's university. There he again pleased his superiors by his ability (for which he thanked his aunt daily) to be as inconspicuous as possible. Nevertheless he had to admit defeat after two years. He could not accustom himself to the academic life. The constant vigilance he had to maintain to keep his more ribald classmates from accidentally killing him for fun sapped his strength. This academic adventure did not visibly harm him but rather cultivated an appropriate earnestness which was the cornerstone in the construction of his exemplary existence.

The father died. This event produced congruous grief in Mr. Van Tussen and corresponding obsequies. The death became quite significant when Mr. Van Tussen was informed that he was the sole heir to his father's modest but comfortable fortune. Joy was marred by the subsequent discovery that, though sole heir, he would have to tolerate a monthly stipend from a separate fund his father had bequeathed to a woman with whom he had enjoyed his old age. The bank's fastidious bookkeeping duly notified Mr. Van Tussen each month. Twelve times a year he was reminded of the cracked image of his impeccable parent. However, even fortunate mistresses die: this one within ten years, of consumption. Mr. Van Tussen never stooped to reproach his father for this eccentricity. He considered it merely a strange whim in a man who otherwise was morally very sound. This filial probity was rewarded by a reading of the old man's last will and testament. Besides pecuniary stipulations, the document was a veritable homily, summarily remembered by its recipient in the phrase: a tranquil life of ethical moderation.

The son followed its advice to the letter. He bought an austere patrician's house on a stagnant canal: a privy of gabled propriety. True to his appreciative nature he furnished it with the asperity of style his aunt had been fond of. His most prized possession was a box in which he kept the testament. The box stood on the mantel above the fireplace between two silver candelabra with tall candles which had never been lit. The ebony had enjoyed such ebullient care that it shone with a satiny luster, darkly mirroring the chandelier. But its decorations made the box particularly interesting. The lid and all four sides had lovingly carved figures of inlaid ivory. Due to his education Mr. Van Tussen was able to recognize the figures.

On the lid were the antic capers of nymphs and satyrs. The realistic detail of Bacchus' satraps surpassed the precision of the sylvan background. The front depicted the birth of Athena. The harnessed goddess was so skillfully carved that she not only seemed to jump from the cleft head of Zeus but also from the box itself. Time had contributed a flaw however: the ivory forehead of Zeus was cracked and a fine fissure ran from his forehead to his chin making the god both frown in pain and smile in relief. An antiquarian had assured Mr. Van Tussen that relics only profit from such accidental imperfections: they are witnesses for authenticity. On the left side of the box was a bold carving of nude Pan chasing fauns through the underbrush. On the right drunk Silenus was giddily examining the enormous flaps of his stomach. These sides were obstructed from view by the massive candelabra. The back side of the box showed Athena wrestling with Hephaestus while on the trampled ground the god's seed grew into a boy. Only at night, when he was quite alone, would Mr. Van Tussen turn the box around.

This antique was the only memory he had of his mother, who had died while giving birth to him. His father had only rarely spoken of his bizarre wife whom he had married in order to consolidate two family businesses. The box was the only personal property in the estate left to Mr. Van Tussen. His mother had kept a packet of letters in it sent to her by a young lieutenant who had died of syphilis in the tropics. A rare visitor could only concur with the devotion and respect of Mr. Van Tussen when he spoke of the box's artistic beauty and its presently edifying contents.

In this celibate house Mr. Van Tussen had regulated his exemplary existence. Each day went according to schedule. The program included a daily stroll through the park. He would leave his house at two o'clock in the afternoon, walk to the park and remain there till half past five. At six o'clock he would eat at his customary restaurant near the entrance of the park, where Albert the headwaiter had served him faithfully the same dinner for over a decade now. When in the park, Mr. Van Tussen would first visit the large artificial lake where he fed the swans and ducks from a brown paper bag. He relaxed in the sun on a bench by the water, watching the graceful commas of the swans' necks punctuating the surface with barely perceptible ripples when they snaked their heads down to catch the soggy, helplessly sinking pieces of stale bread. The stately elegance of the large birds, floating like omens of beauty as if they were dead but carried along by a lurking undertow, could impress him sometimes with a sinister foreboding. This feeling gained reality when on one autumn afternoon of false sunlight he saw a swan break a boy's arm with a blow of its wing. The large white sheet looked so downy while it scythed through the air. The coiling neck with the yellow beak screeched ugly sounds of fury while the reptilian eyes remained obdurate. He had wanted to leave but had not been able to get up. Loud cries finally forced him away and he felt as if he were fleeing the scene of a crime. But there was always the comic relief of the ducks: the bustling comedians of the lake. He liked their prosy colors, their broad friendly beaks and their incessant squabbling about all sorts of domestic nonsense.

After the lake he would stroll through the rose garden, which was the pride of the municipal staff that took care of the park. The roses grew in Euclidean patterns. Their glowing faces were sprinkled over shorn bushes planted with careful consideration for decor. The gravel under his feet was always clean, as if each stone was washed by hand. Despite fall, the gravel was not bothered by dead leaves but lay raked in neat columns which one hardly dared disturb. The greensward grew its perimeter five inches from the low, painted railing; its grass was weekly handtrimmed by the assistant to the chief gardener himself. Forming a perfect circle the grass stopped to delineate the black humus in which the roses were tended. In front of each bush was a freshly painted sign educating the spectator with the flower's vulgar name and its botanical, proper Latin inscription. This garden was Mr. Van Tussen's favorite spot. He would sit on a bench in the pergola where scarcely any noise intruded and where the compelling scent of the roses would drug him with a perfumed narcosis. There he sat as if in a bell of silence where wind and sounds seemed to have been intercepted and made innocuous in order to prevent any blemish on this inscrutable harmony. Usually he would fall asleep almost immediately.

Even in winter (if it didn't snow, rain or sleet) Mr. Van Tussen, fitted into a fur-collared overcoat, lambskin gloves and a cap of otter fur, would hasten to the lake and immediately empty the brown paper bag on the ice. He would walk quickly through the wintry rose garden—dismal and tedious with its skeletal bushes wrapped with burlap in the frozen symmetry—to his usual café, where he would doze over hot punch. Discreetly roused he would leave and fifteen minutes later he'd be eating his dinner at six o'clock in his customary restaurant.

Mr. Van Tussen was contented with his existence. It had maintained a predictable continuity over many years and he could not foresee any likely disturbances. One generally appreciated him because he was innocuous and a gentleman. Older people approved of his manners, while his contemporaries did not speak of him because he never did anything which aroused their interest. Even the cleaning woman, widow De Kraai, liked him. He made no difficult demands of her. He merely expected a spotless house. This task was not too strenuous since he didn't keep animals, a wife or any other creatures. Her employer would make an excellent husband for her encroaching old age, so Mrs. De Kraai did the best she could, without having noticeably progressed during the past decade. She had detected only one oddity in his formidable composure: his addiction to cigarettes. Mr. Van Tussen was the sort of man, she felt, who should smoke either cigars or a pipe, but not those white cylinders which always reminded her of nervous young men. But she had to admit that he even smoked cigarettes with a certain distinction; it was a pleasure watching him slide one from his gold case and light it with the quick flame of his silver lighter.

Mr. Van Tussen was a remarkable man and one would look in vain for someone to speak evil of him. He felt that the admiration he sometimes inspired was totally justified. To himself he admitted only one dissonant: his secret fascination for the swans in the park. He had never relinquished the hope that one of them would repeat that violent scene in the furtive light of an autumn afternoon. Nor had anyone an inkling of his fading dream to become a swan-upper. This was the only employment which ever seriously crossed his mind. He would have gladly exchanged his years at the university for a chance to become such an official in a monarch's or even municipal swannery, where he would have to spend his days catching the birds to mark them with a nick on the beak as either a royal or a commoner swan. In such a position he would finally have been able to trace with his fingers the remex feathers which propel either their flight or their violence. But no one knew these modest desires. Mr. Van Tussen was a truly remarkable man: a man of indelible principles.

The last days of autumn were listless and relentlessly gentle. Summer had already basked its memory in the ruddy bricks of Mr. Van Tussen's house. A spiced autumn breeze vitiated with the crackle of dead branches creased the canal in front of his windows. The trees lining the front of the houses lisped secretively behind their mottled coats in which the green of summer had not yet totally faded. But already occasional rents bared pitilessly the skeletal black branches. The sun was a thick haze over the city and the park. Its docile, sleepy warmth drugged the senses. The wind was oily and sweet. The lidded days yawned with weariness but refused to let go, like aging epicureans satiated with the banquet of the senses. These days would not tolerate any physical exertion, not even irritability. Mr. Van Tussen had to obey this unnatural force of the weather, despite the fact that this had been one day when everything had been just a fraction from the norm. His alarm clock a few minutes late when he got up, his last razor blade dull, the butter turning rancid, the mailman an hour late and then with no mail for him: a day full of such vanishing little nuisances.

He was sitting on the brown, tarred bench by the lake. The swans' bodies were so still that they were indistinguishable from their reflections in the water. Only their slow, long necks were sliding lazily under the surface, barely limning the water which flowed languidly past the white coils. The ducks were silent. The usually so inviting park now seemed to distrust intruders. Mr. Van Tussen sat motionlessly, leaning slightly forward, his hands in his lap. His face was wiped by the retreating sun. It became soft and vulnerable with the sickly sweetness of overripe old men. While watching the hypnotic rhythm of the white pistons he gradually dozed off. He woke with a touch of fear. The sharp little noise of chain against watch embarrassed him a little. He did not notice that his watch had stopped. The park was hung with shadows. Light was a pale ember in the dark. Mr. Van Tussen stood up and the brown paper bag with stale piece of bread fell from his lap. He did not bother to pick it up.

On his way to the restaurant he walked as if he were still half asleep. No one passed him. Not a sound betrayed a presence. He would not have been surprised if the gates had been locked already. He told himself that it was pleasant to be alone like this. It suited him. He was a man who made a virtue out of privacy. He lived with care and precision and avoided the unpleasant mishaps of life which made so many people suffer. No one could change him any more. His choice of life had proved to be correct. No, he wouldn't deny that he had felt some pleasure when he noticed that the widow De Kraai had definite ideas about him. But this indulgence couldn't have lasted very long, lingered for perhaps a few days.

He could always leave her some money without having to go through all the inconveniences of marriage. To start anything like that at his age would be too unsettling. He was a bit too old to change his ways. Solitude had nothing frightening. He smiled while he stooped to pat a dog which had walked for a while next to him. Don't you think I really did quite well? he said while admiring the muscled gait of the animal. I am only patting you. No need to show your teeth. Now really, what's all that about? No reason to be unfriendly to an old man. Maybe not so old yet. I like animals. Swans and ducks. You wouldn't know. You kill rather than feed them. The dog growled softly. Go away then, he said. Go back to your master. He walked on, his pleasant revery tainted with anger because of the dog's rude behavior. He soon forgot the animal and when he reached the restaurant he did not know whether the animal had followed him or not.

When he was almost home his hand touched the rough coat of the dog. He told it to leave him alone. Two blocks further he shouted at it. He made a couple of threatening motions. The animal was not in the least perturbed. It bared its teeth and snarled back. Mr. Van Tussen decided that it would be best not to pay any attention. It would go away. After all, he had never seen it before, he had not fed it anything, he had done nothing to justify its persistence. The more he forgot about it the more its presence infringed upon his consciousness. Ready to climb the stoop to his house he suddenly turned around. It was standing near the curb watching him. He lunged forward and tried to kick it. The dog easily dodged his foot and attacked. Its teeth were long and white. They missed the leg. But the sound of the jaws snapping shut was sufficient to refrain Mr. Van Tussen from attempting any more violence. He tried to persuade it to remain at the foot of the stairs. He told it that it was a real nice dog and that it should go back where it came from. Somebody was missing it right this minute. A little girl was undoubtedly crying for it. He lost his temper again when the animal just stood there and looked at him as if he were some sort of madman. Go away. I don't want you. I didn't tell you to follow me, you rotten mongrel. Damn you, go away, he shouted and ran up the steps two by two. But it took him a while to open the door. He had to search for his keys, find the correct one, fit it into the lock, turn the key, fumble with the knob (it was quite dark by now) and open the door. When he shoved past the door he saw the dog beginning to clear the stone steps with one long leap. He swore again and slammed the door shut just before the dog regained its balance on the slippery marble stoop. He leaned out of breath against the wall of the corridor and heard the animal bark outside with such fierce menace that he shuddered and praised himself for being so quick and so clever.

That night the incident would not leave his mind. He had not been afraid of the dog even though he had certainly realized its size and strength. He had almost patted it again when he had tried to sweet-talk it away. Only when it had tried to follow him inside had he felt a rush of fear. Images of having been bitten flitted through his mind. His throat had been torn open and blood flowed freely down his chest, sucking his life away and leaving his face shriveled like an old apple. He tried to dismiss these ridiculous flutters. Something had to be done about that animal. With this resolution in mind he went to bed and slept a fretful, hostile sleep.

The snow was excessive that winter. It had been freezing for months and the snow had turned into slivered ice. The sun was white and blinding. The sky was a cobalt shield curving away over the horizon. Mr. Van Tussen walked his dog every day. It was nice to have a dog during the winter. It shared his endless nights. He talked to it. He stroked it. He played with it. But mostly he would sit before the fireplace, the dog on the rug, and would stare into the glistening mica of the animal's eyes. These silent conversations were exhausting but rewarding. Dogs were popular pets and nobody paid any attention to Mr. Van Tussen and his dog. People sometimes smiled to show their approval that he had finally found a companion. Yet he still harbored a grudge against the animal for the way it had forced itself upon him. After all, if he wanted a dog so badly he only had to go to the municipal dog pound. He could have strolled leisurely past the cages in which all the varieties of the canine race staged their pitiful show of affection. He would have chosen one after careful thought, blessed by the attendant who would have been happy to have one dog less to bother with and who would have commended him for the excellent choice he had made and what a good deed it was to give a home to one of these poor creatures.

Mr. Van Tussen had to walk the dog every day (no matter rain, snow or sleet) with unrelenting regularity. In the afternoon he would walk the dog through the park. He had to watch the animal's urine bore neat yellow tunnels in the swan-white snow:

steaming tunnels which would sometimes reach the black, cracked soil. Though disgusted by this daily show of uncivilized behavior, he couldn't escape this humiliation for fear that his carpeting and rugs would be ruined. He consented to look upon the dog as a remarkable animal despite these crude necessities. It had a mind all its own. As long as he was gentle and spoke low, as long as he was in a pleasant mood, it would close its eyes and wag the tip of its tail. Yet as soon as he grew irritable or was in a rare bad mood it would become, without any warning signals, suddenly unruly, snarl, bare its teeth, rise slowly, almost as if stretching itself, and move toward him as if to attack. Though Mr. Van Tussen had noticed this he could not suppress his increasingly bad temper.

During the long winter evenings the dog would lie on the rug in front of the fireplace, while Mr. Van Tussen sat in his chair smoking cigarettes. They watched each other with guarded interest. The heat of the fire would scorch the contour of the animal until only a fiery haze suggested its presence. The flames would leap in its eyes and Mr. Van Tussen found an inexhaustible delight following the altering shapes of the fire. But he also watched the eyes to be on his guard against the sudden change from contented dog into dangerous predator. Knowing the reaction of the animal to his behavior he enjoyed surprising the animal by quick shifts of mood. Sometimes he managed to catch it napping. They played these games evening after evening and they became more and more violent. Mr. Van Tussen told the dog that he could not help preferring excitement to the complacently snug silence. Just remember, he would say, that you still don't belong to me. You're a pest, garbage I picked up from the streets and took pity upon. As soon as it is spring I will go to the dog pound and have them take you away. I won't do it yet because it is winter. You would freeze to death in those cages. They don't heat them enough. Of course there are more dogs during these months and more of them are stuffed into one cage so that they can keep each other warm. Dogs are supposed to be clever, as you well know. He would insult the dog, leaning slightly forward in his chair, his hands in his lap, his face erased by the fire to the unhealthy pink of an old man.

He fed meat to the dog. In the beginning he gave it only scraps from the butcher. But he became gradually more and more generous and the dog more and more demanding, until they shared the same meals. Mr. Van Tussen had even stopped eating fish on Friday, which he had always done to honor his aunt's memory.

The dog would tear the meat with quick, sharp jerks of its massive jaws, blood dribbling from its teeth. This daily diversion was utterly fascinating to Mr. Van Tussen. The dog had filled his existence with new interests, feelings and with excitement. He was still contented and felt that he had not sacrificed the regularity of his days to this intrusion. He proudly told himself that he had changed admirably and without effort to a new routine. The only noticeable difference was the fickleness of his moods. Maybe he was finally growing old and (with a smile) growing old as a disagreeable and cranky old man. But this could not be completely true since he, in their games, easily maintained the upper hand.

It was exhilarating to fight the dog on the red rug in front of the fire. Since he liked it so much he had to provoke the dog more and more. He could never recall whether a game had started with anger or pleasure. They would roll on the rug in a ball of snarling fury, snapping and growling at each other. The man could do little more than try to keep the teeth from reaching his throat. But the animal's eyes and dribbling muzzle were always in his vision. The animal pummeled his chest and belly with its paws, tore his chin with dull, hard nails. The man had perfected a surprise attack at the tail. With one hand he would dig into the dog's throat and with the other quickly grab the tail and twist it until the animal yowled with pain. The dog would hurl its weight against the hand choking it, trying to sink its teeth into the man's flesh. His adversary had to let go of the tail in order to protect his face. The animal's furious but steady eyes were always there: eyes in which the flames flickered a wild dance cut by teeth until there was only a white snarling skull through which the fire sparked crystals of blood. In order to end the fight he would struggle to his feet and quickly jump away while caressing and soothing the animal with his voice. It would never attack again but go back to its place on the rug, watching the man intently, watching him with the fire in his eyes. Both would be silent and Mr. Van Tussen would feel the bond of undue friendship between them.

Though it seemed unlikely, the snow began to melt. Winter's fierce majesty slopped into mud. The fights had become more frenzied and Mr. Van Tussen's health began to suffer from the constant strain. Now his chest quickly became a burning cage and his breathing a laborious gasping for air. They still fought on the rug, but Mr. Van Tussen had great difficulty keeping the teeth from his throat. He was no longer able to attack the tail. He needed all the strength he could muster to keep the jaws at bay. More evenings he would sit in his chair without even so much as insulting the animal. He merely sat, muttering from time to time. But the dog had grown accustomed to their games and began to demand that he get out of his chair and start another round. One day it began to snap at his legs to rouse the man to his duty. Its insistence became more and more violent and soon his legs were covered with deep bites. He was finally forced to lock himself in his bedroom while the animal stood in front of the door, growling and snapping at the doorknob. The man had to give in when the animal became too enraged. He would suffer through a fight with such tremendous effort that he needed several days to recuperate. The dog would remain quiet while he tried to regain his strength. But as soon as he felt better it renewed its attacks on the door. Days went by when Mr. Van Tussen could not walk the dog and he knew it was using the house. It was not long before widow De Kraai looked at him with disgust and walked out, saying that she refused to clean up after an animal. Mr. Van Tussen could not persuade her to stay and she left in a huff. He sympathized with her displeasure, but he nevertheless made a mental note to remove her name from his will.

The more torturous the fights became the more Mr. Van Tussen hoped for spring. He promised himself that as soon as spring arrived he would have the dog pound get the animal. He sat in his bedroom with the door locked watching the snow melt. Rains began to wash the streets clean again. Slowly people reappeared. Children's voices called from beneath his window. A new light stole through the sky and some hasty buds shriveled on the trees in front of his house. He even began to think of the swans again. When he had finished with the dog he would apply for the position of swan-upper, without pay. Surely they would accept such an offer. He liked the idea of handling bread again. The dog must be very hungry by now, he thought. It hasn't eaten for a long time. It had been unnaturally quiet and had stopped attacking his door. A splinter of fear began to fester in his mind. Soon he could go outside and get them to take it away. He would leave the room as if he were going to fight, run past it, slam the house door in its face and be on the free, clean street.

But the day when it was clearly spring, the day when he should go, he found himself dizzy and weak. He had not eaten for a long time and he could hardly stand on his legs. He had to support himself by clinging to the walls. His head throbbed and his vision was blurred. It would be outside his door waiting for him. Would he allow himself to be intimidated by an animal? Anger roused him. Fear worked energetically in his body. Soon he found himself hating with an alien but ferocious strength. He took a blanket from his bed which he would throw over the dog to give himself time to get to the house door. He opened the door of his bedroom, looked outside, stepped into the corridor, stumbled and fell over the trailing blanket. He fell with a slight sound, as brittle as a dried leaf. He turned his face toward the living room and coughed on the dust in his dry mouth. Though blinded by daylight he saw the shadow dash from the room in which the logs had long grown cold. When it was upon him he saw that the fire had not left the eyes but roared with flames, burning his head and chest. Its weight pressed him into the crumbling carpeting and he felt the search for his throat. He tasted blood and foam in his mouth. Jerking his head away from the yellow teeth he saw that the door to the street was locked and bolted. Screaming with hatred he clawed himself into his enemy.

THE NEW DELIVERANCE

A Play in One Act

STANISLAW IGNACY WITKIEWICZ

Translated by Daniel Gerould and Jadwiga Kosicka

TRANSLATORS' NOTE: Painter, playwright, novelist, aesthetician, philosopher, and expert on drugs, Stanisław Ignacy Witkiewicz, or Witkacy as he called himself, is Poland's outstanding figure in the avant-garde between the two world wars. Born in 1885, son of a famous critic, Witkacy grew up in the picturesque mountain town of Zakopane in the south of Poland when it was still part of the Austro-Hungarian empire. Educated entirely at home, the boy was encouraged to develop his talents in many directions. By the age of five he was painting and playing the piano; at seven he wrote his first plays, which he printed himself on his own press. Distinguished visitors to the household included his godmother, Helena Modjeska, and the young Sholem Asch. The composer Karol Szymanowski and, later, Arthur Rubinstein were among his close associates.

Copyright © 1973 by Daniel Gerould and Jadwiga Kosicka. All rights, including professional, stock, amateur, motion picture, radio and television broadcasting, recitation and readings, are strictly reserved, and no portion of this play may be performed without written authorization. Inquiries about such rights may be addressed to Toby Cole, 234 West 44th Street, New York 10036.

Destined by his father to be an artist, but uncertain of his direction in life, Witkacy underwent a severe mental crisis in his mid-twenties and was analyzed by a Freudian psychiatrist. After the suicide of his fiancée in 1914, he went with his friend Bronisław Malinowski on the anthropologist's first expedition to Australia. Upon the outbreak of World War I, Witkacy volunteered to go to Russia, where he served as a Tsarist officer and witnessed the last days of St. Petersburg. During the revolution he was elected political commissar by his regiment.

From 1918 until his death in 1939, Witkacy worked in Poland, earning his living as a portrait painter and writing prolifically in many genres. He developed a theory of nonrealistic theater, which he called Pure Form, and wrote over thirty plays, many of which remained unpublished and unperformed. The critics attacked him savagely, and during his lifetime he was regarded as a madman and drug addict because of his experiments with painting under the influence of narcotics. His last years were devoted to philosophy. In poor health and despairing at the fate of civilization, Witkacy committed suicide on September 18, 1939, after first the Germans and then the Russians invaded Poland.

On one of the portraits which he painted in the late 1920s Witkacy wrote the following inscription: "For the posthumous exhibit in 1955." He proved to be remarkably accurate in his prophecy. Forgotten during the Stalin years, he was rediscovered in his native Poland only after the liberalization of 1956, when his works came to play an important role in freeing the arts from Socialist Realism and in the formation of a new Polish avant-garde theater. His twenty-three surviving plays were published in 1962 and, after some vicissitudes with censorship, Witkacy is now recognized as Poland's greatest twentieth-century dramatist, his work translated into fourteen languages and performed throughout the world.

The New Deliverance (1920), one of Witkacy's best-known plays, was first performed under the author's own direction in Zakopane in 1925. More recently, it has been staged by Józef Szajna (co-creator with Jerzy Grotowski of Acropolis) as a forecast of the Nazi holocaust. In New York The New Deliverance was seen off-off Broadway at the CBS Repertory Company in 1973.

Dedicated to Karol Szymanowski

CHARACTERS

FLORESTAN SNAKESNOUT—*thirty years old. Tall, dark blond hair. Completely clean-shaven. Handsome. Dressed in a black jacket and white flannel trousers. Straw hat. Quite elegant. Shirt with violet stripes. Violet tie. Low yellow shoes.*

KING RICHARD III—*hunchbacked. Completely clean-shaven. A small round red cap on his head; his crown on top of it. Wears a breastplate, over it a loose red doublet (or something of the sort) trimmed in black fur. Brown hip boots up past his knees, with enormous spurs. A large two-handed sword by his side.*

TWO MURDERERS—*dressed in black tights. Black masks over their faces. Huge daggers in their hands.*

TATIANA—*graying blonde. About forty years old. Monstrous circles under her eyes. Traces of great beauty and great success with males.*

AMUSETTA—*young girl seventeen years old. Pretty brunette. Light-gray dress with a blue sash around the waist. Blue ribbons in her hair, fanning out on both sides. Black stockings. Black shoes.*

JOANNA SNAKESNOUT—*Florestan's mother. Thin matron, dressed in black with gray hair, but worn the way young women wear their hair; a kind of casque in the middle with the hair wound around. Rather florid face. Sixty-three years old.*

HOUSEKEEPER—*forty-five years old. Ball of grease with rolled-up sleeves. Raspberry-red blouse. Brown skirt. Lemon-yellow apron. Red, sweaty mug. Walks with a waddle.*

UNKNOWN SOMEONE—*dressed like the murderers, but in violet tights and a violet mask.*

SIX THUGS *with instruments of torture—five with beards: two with black beards have pincers; one with a red beard has a saw; two with blond beards have hammers. Dressed like workers, in gray-yellow colors. The one without a beard has a small black handlebar moustache and wears a worn-out black jacket; he has on a red tie and holds a blowtorch, out of which a blue flame can shoot forth with a roar as the need arises.*

(The stage represents a huge hall. Two gothic pillars in the middle. A door to the right in the back, another door also to the right, further downstage, a little behind a dark olive-green sofa which stands to the front of the stage, facing the audience, in a line with the space between the pillars. In front of the sofa, a small yellow table. On either side, two wicker chairs like those used in cheap Italian restaurants. Two more pillars to the right and the left, by the footlights.)

Prologue

Typhon, give me your fire:
My heart I broke on the black edge of the tower,
And out went the lamp
That gleams behind that cage of wire.
In the corner, masked, the murderers lour
Who guard the king
Spewing out his ire.
I enter—daggers behind, windows flung wide,
And the Angel of Darkness flashed through the distant hall.
Enormous mirror, six candles burning by the wall.
PRESIDENT: Oh, poison—I am the torturer.
SOUL: And I am your brother extortioner.
The candles dimmed, the rain-soaked gutters sigh.
Am I here? In the hall?
No, it is fear that chokes me,
"To arms! To arms!" I hear my madness cry.
And I see, I see, most clearly see someone push the door ajar
And six thugs with pincers he before him drives . . .
And in the mirrored wall six thugs
Tear out my sleepless eyes.

When I awoke stillness was all around,
Only the rain-soaked gutters louder sigh.
On the windowpane I see, most clearly see
An ink black spider chase an autmn fly.

(1906)

Scene One

(King Richard III stands to the left, his back up against the pillar, and every so often he makes a move as if he wanted to tear himself away from the pillar and get away. The masked Murderers prevent him from doing so, instantly turning the points of their daggers at him and at the same time producing a loud hiss, made by drawing out the letter: A = Aaaaa! Taking turns, they likewise point their daggers at anyone who tries to come near the King. Tatiana is sitting on the sofa in the left corner, nearer the King, and knits. Amusetta is sitting to the right, that is, on Tatiana's left hand. A lamp is burning overhead. Another lights the half of the hall behind the pillars.)

TATIANA *(knitting, to Amusetta)*. Today I'm going to show you someone new. Florestan is his name. He should be here any minute. I won't tell you anything about him, because I want you to guess for yourself what he's like and what his destiny will be. *(To the Murderers, turning her head slightly, but without looking at them:)* My worthy Murderers, pay attention to your own work like ordinary, hired workmen and don't get distracted by listening to our conversation.

AMUSETTA. I'm so frightfully curious. I had the feeling back then that no one would ever come, that everything was finished and done for in that hideous convent.

TATIANA. Don't be silly, Amusetta dear . . .

(The King roars. Tatiana keeps on knitting. Amusetta remains seated, staring vacantly into space.)

KING. Won't you let me rest for a single moment. Oh! Cursed torturers!! *(The Murderers point their daggers at him.)* My back is growing into this infernal wall, my hump pains me like a gigantic abscess. When will the torment of being rooted to these rough stones finally end?

TATIANA *(to the King, turning toward him and looking at him)*. Five o'clock tea will be coming soon, Richard. You can have a cup of tea with us.

KING *(more calmly, with regret)*. Such strength, and it's all wasted. And yet, if I wanted, not one stone in this entire hovel would be left standing. *(Shrugs his shoulders)* How could they ever come up with something like this? *(With sudden rage)* Really, aren't there any more people left in this world? Have they all become nothing but the wheels of a clock? At least let me wind up the clock!

MURDERERS *(pointing their daggers at him).* Aaaa! Aaaaa!

TATIANA. The clock winds itself. It's one of those self-winding clocks . . .

AMUSETTA *(interrupts her, not paying any attention to the King).* If only he's not one of those common, ordinary party boys. Like all the ones I've been meeting at the balls recently.

TATIANA *(knitting).* Calm down. I'm not one to supply you with inferior goods. But you've got to get him in your clutches right away.

AMUSETTA. But how? I don't know any of the ways yet. They didn't teach us that in the convent.

TATIANA. Your feminine intuition will tell you. He's an extraordinary little rascal. And yet he's much more foolish than you are, even though he's thirteen years older. It's always easiest to get the best of extraordinary people. In daily life they're sheep you can lead by the nose, up and down, and around the barn . . .

(From afar, as though from beyond the walls, the sound of three blows struck on a huge gong.)

KING *(furiously).* A new round of visits is about to begin. Every day to see these hideous dramas of yours, to view your disemboweled psychic guts—along with their contents, to look at your metaphysical navels, stuck on sticks and sold like candied fruits! *(With mounting fury)* Oh! I'll get my hands on life some day! I tell you frankly, this won't last long. I'm not afraid of death, but I'd like to see one more transformation of my role—just once take a walk through the streets of your town. Get out of this cursed hole, the way I once used to leave the Tower.

(The Murderers hiss, with their daggers aimed at him.)

Scene Two

(The same. Enter Florestan through the door in the rear; without taking off his hat, he approaches the sofa from behind. He comes up to the sofa and leans on the back with both hands. Neither Amusetta or Tatiana turns around. Tatiana knits. Amusetta shows signs of nervousness.)

FLORESTAN. Sorry I'm late. I won an important tennis match and I had to change. The contest dragged on too long.

TATIANA *(without stopping her work)*. This is Amusetta whom I've told you so much about.

FLORESTAN. More hideous procuring. But I can make her acquaintance. That won't cost me anything.

(He makes a move as if he wanted to go around the sofa to the right.)

KING. You could say hello to me first!

(Florestan stops and for the first time notices the King.)

FLORESTAN. Where have I seen that hunchbacked monkey before?

(He goes over to the King.)

TATIANA. That's Richard III. He's been given an indefinite leave of absence and now he's doing penance. It's a little comedy that I'm putting on.

FLORESTAN *(going over to the King and taking off his hat, addresses him in an offhand manner)*. I'm very glad to meet your Majesty.

(He tries to draw near and holds out his hand.)

FIRST MURDERER *(pointing his dagger at him)*. Aaaaaaaaa!!!

KING *(to Florestan)*. There's absolutely no need for you to stick your paw out at me, Mr. What's-it, whom I wouldn't venture to call by your true name. You can talk to me from a distance.

(Amusetta looks at Florestan admiringly.)

FLORESTAN. But I say, Madam Tatiana, that Richard of yours is ferocious! I never thought a king could be so coarse and crude. *(The King wants to hurl himself at him and takes a step forward. The Murderers restrain him with the points of their daggers.)* If it weren't for those gentlemen, I'd most likely be lying on the floor chopped in half by that saber.

(He points to the King's sword.)

AMUSETTA *(to Florestan)*. Mr. Florestan, haven't you already spent enough time on that puppet? He's not a real person . . .

KING *(furious, interrupts her)*. I'm more real than all those characters of yours that I've had to watch here for the past five years. I'm like a barrel of nitroglycerin. When I finally deign to explode, not even the dust from these dungeons will remain.

(Florestan listens to him with a polite smile.)

AMUSETTA *(not paying the slightest attention to the King)*. That doesn't matter. Mr. Florey, come over here this minute, please, next to me.

FLORESTAN *(looking at her for the first time)*. But what a lovely girl. *(Approaching her)* I just finished a story today in which there's a character exactly like you.

TATIANA *(interrupting him)*. Mr. Florestan Snakesnout.

(Florestan greets Amusetta and sits down on the chair to the right, his left profile to the audience. He's just about to say something when the King interrupts him.)

KING *(in earnest)*. That is how I condense my strength. What a pleasure to crush people when one is young . . .

TATIANA *(to the King)*. Richard, we've had quite enough of your effusions for now. Calm down and think about teatime coming up next.

FLORESTAN *(to Amusetta)*. Miss, I see that Madam Tatiana hasn't made any mistakes about my taste this time. As a matter of fact, I needed a new woman. You know, people think I'm a Don Juan. It's not true. I'm only concerned with certain fleeting states which I subsequently use for artistic purposes.

AMUSETTA. But you're not an artist, are you? That's so common now. Everyone's an artist these days.

FLORESTAN. I despise art, Miss. I use it the way I use all other amusements, which I also despise. The only plaything I don't despise is woman.

KING *(angrily)*. I can't stand much more of that scallywag's company!

FLORESTAN *(to the King)*. Quiet, you old marionette! *(To Amusetta)* Go on, I'm listening. What did you want to say?

AMUSETTA. Then what are you really? You're starting to become extremely mysterious.

(Tatiana looks at Florestan with a smile.)

FLORESTAN. Even if I told you what my profession is, you wouldn't believe me anyhow. I'd rather not say. To the extent that I am able—I am everything.

AMUSETTA. I have the feeling you've got a frightful appetite. You impress me as being a land-shark.

KING. Why not say: a simple swine. Can't you see he's an ordinary pragmatist?

AMUSETTA *(turns to the King for the first time)*. And just what is a pragmatist?

(Florestan tries to interrupt her, but she motions him to be quiet with her hand and listens to what the King says.)

KING. A pragmatist is an ordinary beast, with only this difference, that by making a theory out of his beastliness, he tries to convince others that it's the only possible philosophy.

FLORESTAN. He's raving. First of all, I'm not a pragmatist, and second, pragmatism isn't what the King imagines it is. As for philosophy, I'm the creator of a school of transfinite monadologists. We totally reject the idea of dead matter.

AMUSETTA. Well, fine—then this table can fall in love with me?

FLORESTAN. Not the table. But the living beings that make it up can be in love with each other, the way we can—we're just another small particle of the universe like them. You have the eyes of a coral viper. I remember how once one of them slept with me, on my breast while I was asleep too, and when I woke up we looked into each other's eyes.

AMUSETTA. Then what happened? Didn't it bite you?

FLORESTAN. If it had, I wouldn't be here talking with you now. A second later it was breathing its last between my fingers, as its tail kept hitting me across the arm and even across the face.

KING. Served you right. The only thing to do with people like you is hit them right in the snout. Oh! If I could let you have it just once! *(He makes a move forward. The Murderers restrain him with a hiss.)*

FLORESTAN *(to the King, coldly)*. I'd advise your Majesty to stop making light of my person. Some others have long remembered making a few little sarcastic remarks like that.

AMUSETTA. Are you strong? Have people often made fun of you throughout your life?

FLORESTAN *(flexing the muscles in his right arm)*. Just feel my biceps.

AMUSETTA. But it makes me somehow feel embarrassed. I have the impression it must hurt when it's so hard.

FLORESTAN. There's never been a man yet who dared make light of me. I'm not counting that madman. *(He points to the King.)*

KING *(laughs savagely)*. Ha, ha, ha! For the time being there's nothing else I can do except laugh. But we'll see yet. I believe that in the last analysis everything must make some kind of sense in the general scheme of life.

(The gong sounds three times.)

Scene Three

(The same. Enter the Housekeeper carrying the tea things and some cakes on a tray. She goes over to the table and arranges everything on it.)

FLORESTAN. Our conversation's been interrupted. I can't say that I've fallen in love with you. In any case you're not a part of the background, but a separate complex of very intricate combinations. As such you exist in my consciousness. That's a great deal, a very great deal.

AMUSETTA. Complex is a horrible word. It sounds like a compress that got flexed. In the old days women used to be told they were like flowers. I read about it in the convent after lights out. Just what is this background of yours?

(Florestan tries to answer. The King interrupts him.)

KING *(to Amusetta)*. Undoubtedly hideous swinishness. You shine like a wonderful black diamond encrusted in a piece of rotten headcheese. *(To the Housekeeper)* And how are you, my love? Do you still believe me, despite my monstrous lies and my no less revolting infidelities?

HOUSEKEEPER *(respectfully, finishing setting the tea things)*. I am your humble servant, Your Majesty.

FIRST MURDERER *(to the Housekeeper in a rasping voice)*. What's the time?

HOUSEKEEPER. Ten at night.

SECOND MURDERER *(in the same kind of voice as the First Murderer, lowering his dagger)*. We can go get some sleep.

(They both lie down and fall asleep against the wall. In a moment they can be heard snoring.)

KING *(straightens up with relief)*. At last! Bring on the tea!

(Stretching himself, he goes over to the table. Exit the Housekeeper.)

TATIANA *(putting down her knitting)*. Mr. Richard York—Mr. Florestan Snakesnout.

(Florestan gets up. They shake hands. The King sits down in the chair to the left. He unbuckles his sword and leans it against the sofa.)

KING. You'll forgive my remarks, Mr. Florestan. When I stand there against the wall, my nerves are always on edge.

(Tatiana pours the tea.)

FLORESTAN. But exactly why do you stand there?

TATIANA. Don't be indiscreet, Mr. Florestan. Go on amusing Amusetta.

FLORESTAN. It never even occurred to me that there was any kind of mystery connected with it. We men of action, men whose lives are all creativity, do not acknowledge that there are any mysteries. Existence in and of itself isn't any more mysterious than this cup of tea. I drink it and digest it. It acts agreeably on my nerves. What is there beyond that? Just tell me. Nonsense.

KING. Yes—but what is there beyond nonsense, Mr. Snake . . .

TATIANA. . . . snout.

KING. Oh, that's right, snout. Look, my paid murderers are asleep over there. They aim at my heart until the appointed time comes. Don't we all have our own murderers?

FLORESTAN. But the ones you have here are real. You're talking symbolically.

KING. Just try to wake them up. Go over there and see how real they are.

FLORESTAN *(getting up)*. No—that's impossible.

(He goes over to the Murderers and shakes them).

TATIANA *(to the King)*. How can you make fun of that poor, unfortunate great man?

KING. Let that teach him, the scum. I went through a lot before acquiring the knowledge of life I have now.

AMUSETTA *(to Tatiana)*. Then he is a great man?

TATIANA *(to Amusetta)*. You'll find that out for yourself.

(Florestan comes back. He has an odd look on his face.)

KING. Well, what have you got to say now, Mr. Snakesnout?

FLORESTAN *(controlling himself with difficulty)*. Nothing to it—these are little known symptoms of hypnotism or something like that.

KING *(ironically)*. Hypnotism!! *(Emphatically)* Those are corpses, my friend, corpses that snore. How do you know I'm not a corpse too?

FLORESTAN *(disconcerted, tries to shrug the matter off)*. The most courageous people have their little weaknesses. I knew a certain cavalry officer who would charge artillery with a detachment of horse, but who quaked at the sight of a rat or a mouse. I have known cavaliers who were afraid of a white spot on a wall. Do you know those nights in the south of France when everything white shines phosphorescently and the darkness seems to be full of phantoms?

(His voice betrays that he is on the verge of tears.)

KING *(banging his fist down on the table so hard that all the cups jump).* That's enough of that rubbish!! You are an utter coward!!

TATIANA *(gently).* Calm down, Richard. Not everyone has your powers of endurance in the face of the strangeness of life. *(To Florestan)* We live in dark corners and byways. Life in the grand manner, out there in the fresh air, has been poisoned by the weakness of the debased mob. We can do nothing, we're like those fish that burst when they're pulled up out of the unplumbed depths . . .

AMUSETTA *(to Florestan).* Still you're not so strong as you said you were a moment ago. I'd so like to believe in someone's strength, in boundless courage, in the possibility of overcoming the pain or fear caused by the mystery of the other world!

FLORESTAN *(in an artificially calm voice).* That's what we have minds for. Animal reflexes exist in man. The tiger is courageous, but does it deserve any credit for that? We're all socialized animals . . .

KING. Not all of us, Mr. Snakesnout.

FLORESTAN *(not paying any attention to his words).* It's only by using our minds that we can defend ourselves against certain primitive reflexes which at any given moment can gain the upper hand. Not over us—of course—only over our nerve tissue.

AMUSETTA *(to Florestan).* You're beautiful, that's something positive. But when you talk, I have the impression I'm in a class at the convent . . .

KING *(to Amusetta).* Yes, my child. Admit that, if it weren't for my advanced age, you'd prefer me to that decadent young whippersnapper.

AMUSETTA. Your Majesty, I think you're right.

FLORESTAN. You'll forgive me, Miss: I'm not an old man. I don't fool around with the various little tricks that people anxious to believe in artificial mysteries use to deceive themselves. I am, first and foremost, a contemporary man. Those old beliefs, those attempts at hypnotism—they're bugbears to scare children. We have two things which the people in ancient times were lacking: thought and organization.

AMUSETTA. Yes, but that's not beautiful. You're not an artist so you don't understand that . . .

FLORESTAN. Miss Amusetta, I can be everything. I'm going to tell you who I am—come what may. I am the vice-president of the largest metallurgical works in the world, but when I want to, I paint so well that none of the cubists is in the same class with me. *(To the King)* I know what you're going to say, Mr. York. I never studied drawing—that's true. But I paint better than all of them. I sit down at the piano and I play as well as they do, those great musicians of ours. And maybe even better. By pure chance I create such combinations of sounds that the most sophisticated connoisseurs of music are bowled over by them. Everyone tells me: Why don't you exhibit? Why don't you publish? For the moment I don't want to, I write as well. Oh, if you knew what I created today! All the creative writers will be green with envy. *(Proudly)* There's just one thing: they take it seriously, but I am a vice-president—that's my business, and all the rest of it is amusement; art, women . . . oh, excuse me, Miss, you alone are different.

AMUSETTA. Yes. You say the same thing to every girl you meet.

TATIANA *(significantly)*. Believe me, he doesn't. I know him well.

KING *(gobbling up the petits fours)*. Well said. I'm glad to see that there's at least one scum who speaks the truth. In the name of God we slaughtered one another like butchers. But that God was something for us. For you pragmatists—because you are a pragmatist, Mr. Straightsnout—God is just nonsense in which you can believe when it brings a suitable rate of profit.

AMUSETTA *(to Florestan)*. Mr. Florestan, is that possible? Are you really what the King says you are?

FLORESTAN. But really, Miss Amusetta! He has a primitive brain incapable of grasping things as intricate as our complicated ways of thinking, I say: our ways, our contemporary ways of thinking. That man has a head like a double bass and a hump that's even bigger. He's degenerate in terms of his own times, but he doesn't have the faintest idea about the precision of thought that we have reached. That's the intellectual shorthand of the Papuans. Go to New Guinea, Mr. York. You'll find worthy disciples there . . .

KING *(his mouth full of petits fours)*. Tatiana—let me—I can't stand it any more. Let me teach that young whippersnapper a lesson. In our style. Like the Yorks and the Lancasters—damn it all—in our style.

(He raises his hand. Tatiana grabs him by the arm.)

TATIANA *(confidentially)*. Don't bother, Richard. I'll take care of him myself. It'll all work out by itself. That silly little goose

(Points to Amusetta) is the litmus paper by means of which we'll get to know his works—what do you call them—the fruits by which . . .

KING. Oh, that's it, that's it, the fruits. Mr. Straightsnout, where are your fruits?

AMUSETTA. I only see dried prunes, candied papayas, the apple Eve gave Adam in the form of glazed sweets, but I don't see any fruits . . .

(Three blows on the gong)

TATIANA *(anxiously)*. What's that? No one else was supposed to come tonight.

(Waiting)

Scene Four

(The same. Enter Joanna Snakesnout through the far door; with a rapid step she goes over to those present along the right side. They all look around.)

JOANNA. Where is my son? Where is Florestan? I haven't seen him for ten years. I heard that he'd fallen into bad company. Are you the ones who are tempting his noble soul? He was always noble!

FLORESTAN *(getting up)*. Mother! That's not true. This is a den of evil spirits. Over there they're torturing some crazy old man who keeps insulting everyone in the grossest terms.

JOANNA *(bursting out)*. It's you! It's you! How big you've grown! *(Runs around the sofa to the right and throws herself at Florestan)* Is it really you?!

FLORESTAN *(controls himself)*. It's me. It really is. *(Impatiently)* Mother! Calm down. There are strangers here. I don't have anything to do with them! *I* am a vice-president. It's by pure chance . . .

JOANNA *(in Florestan's arms)*. You're mine! My one and only! At last, after so many years of torture . . . *(Notices Tatiana)* You here? You whom I practically brought up? You dare drag my son into this filthy cesspool?

TATIANA. Joanna! There are strangers here. Couldn't we put off settling our personal accounts till later?

(The King laughs savagely.)

JOANNA *(ironically)*. Personal accounts! Don't you understand that he's my son, my darling Florey? Who is he? Has he finally become someone? My dearest son! Tell me who you are! Tell me quickly! I have a weak heart; I'm dying of anxiety. Have you been untrue to yourself, you're keeping so terribly still and your eyes are wild like some madman's from another world?

(Florestan stands with his back to the audience. The King and Amusetta watch this scene with genuine interest.)

FLORESTAN *(insincerely, with feigned emotion).* Mother! My dearest mama! Everything will be cleared up. I'm the same as always. Save me! They're desecrating my most sacred ideals!

JOANNA. You dare ask me to save you? You?! After ten years of separation? You dare? That's horrible. I've gone astray myself. I want to escape from my own thoughts too. *(Terribly disillusioned)* I expect to be saved by you, and you, you. . . Oh, you're as vile as ever. You're a vampire who's lived off my heart all your life and sucked this poor abandoned heart of mine like a hideous bloodsucker!

(Breath fails her, she clutches at her heart. Amusetta springs up and supports her. The King gets up and bangs his fist down on the table.)

KING. Shut up! Cackling geese! Now you're going to listen to me. *(To Joanna)* Don't you recognize me?

(A pause)

JOANNA *(after a moment, leaning on Amusetta).* Richard! Is it you?

KING. Yes—it's me. I didn't know you had such a son. He's a zombie, not a human being. He's a soulless automaton, not a man. He's the poisonous spittle spewed out by your rotten society! He's—I don't know what, words fail me. That seducer of silly bleating sheep, that zero, that double zero . . .

JOANNA. Shut up—I love him!!

KING. Fortunately he's not my son. At last I know your real name. He's the son of the monstrous family of Snakesnouts whose sole representative I see here before me. Thanks to him I now can appreciate your entire wretchedness at the time you became my mistress, you thrice-cursed, common stupid hen! He is a Snakesnout! *(In disgust)* It's that foul blood which now reveals to me the secret in our relationship, the secret of why you were never really mine . . .

TATIANA. So that's how it was? So Richard was your lover? So you lied to me? Very well. I'll have my revenge too. *(To Florestan)*

Florey! Do you remember that night, that night when you abandoned me forever? Can you ever forget those diabolical sensual pleasures that have devoured you ever since, from which you are running away, but from which you can never escape. I've taken my vengeance for those other women! That's why I gave you all those others, because I knew that you'd deceive them the way you deceived me then.

KING. Oh, I like that. That's in our style. That I understand. There's strength there.

(Joanna, half dead with pain, keeps leaning on Amusetta, who looks at each one of them in turn without understanding a thing.)

FLORESTAN *(to Tatiana in a broken voice)*. Don't talk like that. My whole life is worth absolutely nothing on that account.

TATIANA *(to Florestan)*. You loved me, and you still love me and only me. *(To Joanna)* Look, you wretch! He loved you with the hideous, selfish love of a spoiled megalomaniac. But his body belongs to me alone, whenever I want it. And through his body I'll have his vile soul, whenever I want it too. I didn't want it, because he was too disgusting for me in his degradation when he begged me for mercy like a hideous worm. Now watch. *(To Florestan)* Tonight I'll be yours. There is nothing in you except this desire. There's nothing ahead of you in life, because you're just one huge lie, and that lie is me, me! I alone enslaved your vile soul. I used you the way people use inanimate objects. You are nothing. You're my plaything and that's all you'll ever be, even if I live to a hundred and twenty.

JOANNA *(weakly)*. Florey, tell her that it's a lie . . .

(A pause)

FLORESTAN *(in despair)*. No, I can't tell her that. I'm what Tatiana thinks I am. I love her. Her and her alone, because she is a lie that reaches to the height of my own lies. My lies are innumerable because I'm a man. She unites them and sanctifies them in one huge orgy of falsehood. I don't have a mother any more and I never had one. Today I started to love that silly little goose *(Points to Amusetta)* and what is that actually but one of my little lies from the past, the kind I used to deceive even you, mother dear! It was your own fault and you will be punished for it.

(Joanna collapses to the floor with a scream. Amusetta puts her on the sofa. The King runs over to Joanna, examines her, and checks her pulse and breathing.)

TATIANA *(joyfully)*. For the first time in years I really feel I'm living. Oh, no one will ever know how much I've suffered!

KING *(to Florestan)*. You've killed your mother, young man. Even I wouldn't have been capable of that.

(Amusetta cries and wipes her eyes and nose with a handkerchief.)

TATIANA. That's not true, Richard! She killed herself. Remember your own youth.

KING *(in a conciliatory tone)*. Oh, yes, that's right. No one is perfect. Of course . . .

TATIANA *(interrupts him)*. Florey! You're mine! Come, humble yourself before me in the presence of the King. *(The Murderers wake up and stretch, then sit down on the floor. Florestan falls down on his face in front of Tatiana. She stands beside him and puts her foot on his head.)* Do you feel it now? Do you feel the new life entering you at the ultimate destruction? The meaning of your life lay in destruction alone and you have built the foundation for your splendid edifices on a frightful quagmire that swallows up every living thing. Come and acknowledge the sole truth of your life.

KING *(slowly backing over to the pillar to the left)*. Can't you all see that none of this is what it seems? It's only make believe. The only truth is that my hump is growing into this wall.

TATIANA *(to Florestan)*. Get up and come with me. Perhaps when you've really destroyed your life, the truth will grow up out of your hideous male lying.

KING *(leaning against the pillar, the Murderers standing close by him)*. Everything is sham.

(He is interrupted by three blows on the gong. The door to the right, nearer the audience, opens and the Unknown Someone in violet tights and a mask shoves a bunch of Thugs into the hall. The light in the other part of the hall goes out and the front of the stage is thrown into much brighter relief against the background of the dark interior.)

AMUSETTA *(frightened, to Tatiana)*. Who's that? What do those people want?

TATIANA *(taking her foot off Florestan's head; uneasily)*. I don't know. We'll find out in a minute.

KING. At last I see some of our kind. They'll give him a real working over.

(The silent group of Thugs slowly emerges from the shadows,

pushed forward by the Unknown Someone. Three blows on the gong. The door at the back opens. The Housekeeper comes out of the darkness slowly. She draws near and leans against the back of the sofa.)

HOUSEKEEPER. Well, now there'll really be something amusing.

(Tatiana goes over to the bunch of Thugs, who slowly draw near from the right-hand side. Florestan remains lying on the floor, face down, without moving. The Murderers stand in front of the King with their daggers aimed at him.)

TATIANA *(to the Thugs)*. Gentlemen, there's a dead woman here. You can't do anything here today . . .

UNKNOWN SOMEONE *(in a high-pitched, almost feminine voice)*. That's no concern of ours! Don't meddle in other people's affairs!

(The Thugs move forward to the front of the stage and surround Florestan's prone body.)

FLORESTAN *(without getting up off the floor, in a weak moan)*. Mama! Save me! I didn't do it! They did, those awful people . . .

KING *(with sudden fury, trying to throw himself headlong)*. Oh, that clown is driving me over the brink!! *(To the Murderers)* Let me go for a minute only, just one little minute!!

(Tatiana stands to the right of the sofa, hugging Amusetta in her arms.)

HOUSEKEEPER *(to the Murderers)*. Let him go! Let him have his fun for once.

(The Murderers step aside and let the King pass; he goes over to the sofa and takes his sword, which he had forgotten to buckle on.)

AMUSETTA *(whimpering slightly)*. And what's going to happen now? Oh, why did I ever come here! It's all so appalling. What's he lying there for? Who are those people?

(The Thugs kneel in a circle around Florestan, with the exception of the Unknown Someone. The clean-shaven Thug starts up the blowtorch out of which a violet flame shoots with a roar.)

TATIANA. I don't know myself. It's as much of a surprise to me . . .

KING *(standing on the left side of the Thugs, draws his sword out of its scabbard)*. Step aside right this minute!! I'm going to settle the score with him for everything and for all of you too!

UNKNOWN SOMEONE *(who has been standing next to the Thugs, facing the audience)*. Your Majesty! *(Goes over to the King and speaks coldly)* To your place this minute, if you please! Over there!

(He points to the pillar.)

TATIANA *(to the King)* Richard! Defend him! I'll be yours till death, only let him live!
FLORESTAN *(stretched out on the floor).* Hurry up! Just get this agony over with!
(The King retreats before the Unknown Someone and stands against the pillar once more, his drawn sword in his hand. The Murderers stand next to him.)
AMUSETTA *(to Tatiana).* Why doesn't he defend him?
UNKNOWN SOMEONE *(from the left side to Amusetta).* Evidently he can't. *(Emphatically)* Do you understand, Miss Amusetta, he can't. *(Suddenly to the Thugs in a shrill voice)* Get him! Get him!
(The Thugs, kneeling, suddenly bend over Florestan and begin torturing him, each in his own way: with pincers, hammers and the blowtorch. A terrible scream from Florestan is heard.)
KING *(suddenly strikes one of the Murderers, on his left hand, in the teeth with the hilt of his sword, hits the other one [to his right] on the head with the blade, leaping into the place where the first one was and then dashes forward with a yell; the Thugs stop their torturing).* Oh! Enough of that vile screeching! That degenerate doesn't even know how to suffer!!
(He finds himself eye to eye with the Unknown Someone in violet, who gently takes his sword out of his hand.)
UNKNOWN SOMEONE *(pointing with the sword to the murky depths of the hall).* There is the way for your Majesty!
(The King retreats and goes to the left, circling the sofa, then heads for the door to the right in the rear. His heavy footsteps on the flagstones and the jangling of his spurs can be heard.)
HOUSEKEEPER *(turns to watch him as he goes).* Good-bye to Your Majesty!
AMUSETTA *(cuddling up to Tatiana).* Why didn't he defend him? What's going on?
TATIANA. Quiet—quiet. This is the way it has to be—this is the way it ought to be—this is the way it will always be . . .
(The King leaves, slamming the door furiously.)
UNKNOWN SOMEONE *(to the Thugs).* Get on with it, gentlemen! Do your duty!
(The Thugs throw themselves at Florestan once more and start torturing him again. Florestan roars with pain.)

CURTAIN

3/5/20

LIFE UNIFORMS

A Study in Ecstasy

WALTER ABISH

I've come to depend on Arthur more than I care to admit. Almost daily we discuss the most recent disaster. Almost daily we also discuss Mildred. I enjoy talking about Mildred. Arthur has a probing mind. A relentless mind. He is not convinced that I understand Mildred. He wishes to broaden my understanding. In a sense the people he brings to the apartment are there for that purpose. To broaden my understanding. Arthur has also mapped most of the unsafe buildings in the business district for my benefit. I do my best to avoid them. Some weeks as many as three buildings collapse in a single day. Because of the alarming rate at which buildings are caving in, it may take a year or longer before the rubble is cleared away and the site is turned into another parking lot, an amusement park, or simply left vacant. Still, the high mounds of rubble have little effect on the people. Life seems unchanged. There is, however, a certain broadening understanding since Arthur was hired by one of the city agencies to document these disasters. By now, determining which building will disintegrate next has turned into a fine science. Some of the best minds sit in a room in City Hall and feed facts into a computer. Once agreement has been reached on a site, Arthur is sent to photograph it. He uses an old German camera. Sometimes the camera is mounted on a tripod. Quite a few of the photographs Arthur takes make the front page of *The New York Times*. It is a grim business.

I met Arthur as he was focusing his Leica on the building I was about to enter. I hadn't noticed the thin cracks in the wall . . . those ominous cracks. Hold it, he shouted. Somehow the urgent note in his voice made me stop in my tracks. Less than a minute later the fourteen-story building caved in. Almost gently, floor by floor, it lowered itself to the ground, while raising a huge cloud of dust. A few people jumped out of windows. Arthur kept taking photographs. I never interfere with the larger scope of things, he later explained. You stopped me in time, I reminded him. But I recognized your face, he said.

The people at the electric company do not visit the disaster sites. They have no need to. They can tell what happened from merely looking at their instrument panel. They can determine the number of kilowatt-paying customers they lost on each floor, as the needles on the instrument panel flicker. There's much to be said for electricity.
I carefully screwed the 200-watt lightbulb into the socket. I did it on the advice of Arthur. I did it also to better illuminate what I was doing. In the past year Arthur had lived in eight different locations. That kind of experience is not to be dismissed lightly. Structurally, he said, the building in which I was staying seemed sound. He pounded on the floor with a stick to demonstrate its soundness. However, the elevator was too slow in responding, and the lighting was inadequate. Personally, being aware of the trouble the electric company had taken to bring electricity to this building, and above all to this apartment, I was less critical of their somewhat sluggish performance. After all, the work had entailed laying miles and miles of electric cables, most of them underground. Hundreds and hundreds of man hours. I kept thinking of those hundreds of man hours spent laying cables. All the same, following Arthur's advice, I did not inform them that I had screwed a 200-watt lightbulb into the fixture. It was quite conceivable that the socket was not made to carry anything as large as a 200-watt lightbulb. There may be something written on the socket to that effect. The man who comes to read the meter was here this morning. My heart was pounding as he stood beneath the 200-watt lightbulb. He glanced fleetingly at the most recent enlargement of

Mildred. Except for the pile of photographs on the table, the table was bare, since I was afraid that any additional weight might be too much for it to support. How did I arrive at such a conclusion. Years of experience, that's how. The meter man briefly rested his flashlight and ledger in which he entered the kilowatt-hours on a conveniently closeby chair, and studied the ecstatic expression on the topmost photograph. I had not invited him to do so. But given the bareness of the table, the photograph on top of the pile was extremely conspicuous. Gravely, for a few seconds the meter man focused his full attention on the ecstatic expression of Mildred. He struggled with some undefinable emotion, but finally left rather abruptly without saying a word. I can understand his difficulty. His training had not prepared him for such an eventuality.

Everyone who has seen Mildred under the glare of a 200-watt lightbulb has found the information Arthur has been able to elicit from her most appealing. Everyone was quick to agree that Mildred had a pair of marvelous legs. White sensuous legs. Needless to say, Mildred is not completely unaware of the effect her legs are having on some of the occupants of this apartment. Similarly, I was given to understand that her husband, Mr. E. Batch, is not totally unaware of their effect either. However, he doesn't have a single 200-watt lightbulb burning in their home. This deficiency is further aggravated by his lack of awareness regarding the existence of this remarkable group of photographs. When he does hear of it, it will doubtlessly come as somewhat of a surprise. Still he will be gratified that none other than his wife Mildred had been chosen for this study.

Mr. Batch works in an office in the city. From his desk he can see an occasional cloud of dust envelop a tall building. Much time is wasted in the office by everyone standing at a window watching the tall and by now familiar-shaped columns of dust that after an hour subside, revealing a fresh vacuum. Mr. Batch believes everything Mildred tells him. Perhaps he believes her out of a fear of what might happen if he failed to do so. But that is hypothetical. He believes that Mildred has a close friend named Paula who lives in an apartment building that is located less than ten blocks away. He doesn't know Paula's second name. He has never bothered to inquire. Mildred, I believe, married Mr. Batch because he is industrious and because he does not ask too many questions. His questions are reserved for dinner. Why aren't we having any veal cutlets, he might ask her. Although it does strain his credulity, after all he is not a complete ass, Mr. Batch believes that his wife visits her friend Paula at least four times a week. You are seeing a lot of her, aren't you, he once remarked. In the evening Mildred gets on the phone to Paula, and in his presence refers at great length to her visit that afternoon. Of course Mr. Batch cannot be certain that it is Paula on the other end of the line. Even if he spoke to Paula he could not be certain since he's never met her. What is she like, he once asked Mildred. Paula loves anything to do with literature. Paula is dark and has a Slavic accent, said Mildred firmly. Mildred's firmness is a wall erected by her to prevent Mr. Batch from casually exploring the inner recesses of her daily pleasure.

In the electric company an employee jots down on a chart that I have broken a company rule by screwing a 200-watt lightbulb into an empty socket that was not designed for 200-watt lightbulbs. It will be some time before the electric company can take any action. They have other concerns, other priorities that will delay any action they may take. Although, in a sense, one can say that the action has already begun with the notation on the chart. The coded notation, by itself, appears perfectly harmless. It may be nothing more than a checkmark. The absence of windows in the electric company prevents the employees from seeing the rising and subsiding clouds of dust. Everyone has more time to concentrate on the paperwork. The information that exists on paper is not necessarily conveyed by word of mouth, but carried by a messenger. He carries the checkmark to a secretary. The secretary who is seated at a gleaming Formica table doesn't know my name. She types it without properly looking at it. On the windowless walls are large calendars with reproductions of attractive foreign landscapes. Nowhere is there the slightest indication of a cloud of dust.

The secretary who is typing my name is using an IBM electric typewriter with an 18-inch carriage. She sits on a swivel chair. Industriously she is utilizing the electric typewriter with the full cooperation of the electric company. They even know when her slender and quite graceful fingers are pressing the letters on the keyboard, but they do not know what letter she is pressing at any given moment. If she were to press the same letter over and over again, none of the men who are responsible for feeding the electricity into the wires would be the wiser. They are not oblivious of her fingers. Her fingers have a certain hold over them. It is an innately satisfying hold. Like most men they are receptive to certain letters. One might say that they respond more quickly to specific letters. In most instances they prefer the letter F. Fruitful, fancy, and forlorn are only three of the several hundred combinations racing through their minds. Still it can be said that their F-preference is a common preference. It is, one thing remains sure, a preference uniting the men who splice the cables, and the man who reads the meter, and the elevator operator in my building. All are practitioners of F. F is a factor in their life, as it is in mine. All the same, unlike them I never use the elevator because it is unsafe, and one rides up and down at one's own risk. These are the words of the management. As the elevator rises the risk increases.

Who are all these men, asks Mildred as she enters the apartment. She sits on my chair without once thinking that at some point in time the legs might give way. She lies on my couch with the same disregard for danger. The letter F rears itself in her head as well. Frequently she substitutes another letter for F out of a misplaced sense of propriety. Yet the letter F is visible on the enlargements. It is visible when she crosses her legs, which she does frequently.

What does Mildred do when she's not visiting us, asks Arthur.

Arthur secretly uses an electric shaver. To confuse the electric company, he switches off every thirty seconds. Since it is not my shaver, I've not informed the electric company about it. They have their doubts. They have gone over the list of all my electrical appliances. Those thirty-second spurts of electrical activity continue to confound them. They cannot determine what is causing them.

I have lived all my life in the city. Most worthwhile efforts take place in the city. An indescribable alertness divides the people who live in the city from the people who live in the country. Mildred is a city person. She takes the elevator to the sixth floor. The absence of large tracts of freshly plowed earth does not upset her.

Who are all those men in your apartment, asked Mildred breathlessly. Who are all those men? They're friends of Arthur, I said. When your hand touches me, she said, I break out in goose pimples all over. I took a certain satisfaction in hearing that. I knew that Arthur and his friends must have heard it too.

When I'm not working in the darkroom I am examining my next move. I know my next move may well depend on those guys at the electric company. What made Mildred say: As soon as you opened the door, I wanted to glide into your arms. Is that statement consistent with what I see on the enlargements. I enlarge

Mildred. It is work that fills me with a private satisfaction. It is quite gratifying. Enlarging Mildred the better to see her F stop. Everyone who enters stares expectantly at the enlargements on the table. They seem to know intuitively that the woman in each of the photographs is Mildred. They seem to know that the slight smudge on one was caused by my hand. This hand has caressed every inch of the eleven-by-fourteen paper, I tell Arthur and his friends. In the darkroom this hand is always protected by a rubber glove. The lengths of the exposures are listed in my notebook. The notes represent years of study. They represent a boundless energy and an acute vision.

The doors on either side of my desk are always open. I'm inclined to believe that they can no longer be closed. But I don't wish to subject them to that test. Open as they are, they disclose another brightly illuminated space. The threshold I have noted in my notebook functions as a kind of boundary or frame. Sometimes I cross that frame. Sometimes I find a reason to cross the boundary. Had the door not been opened I would have been compelled to knock or to rattle the doorknob. When Mildred lies down on the couch in the adjacent room I can observe her from where I am sitting. I cannot see her when I am in the darkroom. She knows this. Arthur and the others know this. No one has ever tried to close the door in my face. It would be inconsiderate, this, after all, being my apartment.

I received the invitation to lecture at the Felt Forum by mail. The invitation, couched in the appropriate polite form, was contained in a legal-sized envelope. Most messages arrive in envelopes. Most are dropped in my mailbox downstairs. If I scrutinize the mailman when he delivers the mail, it is not done to embarrass him. It is done to elicit some kind of information. . . The mailman once delivered a package to my door. I did not invite him in. Whenever I look at him I feel that he has not forgiven me for not inviting him into my apartment and showing him Mildred's photographs.

If I am constantly conducting myself with a certain indisputable authority it is because I am describing a phenomenon that is still new. In time the newness will wear off. In the next room a woman is undressing. I can see her quite clearly. Her every gesture is studied. She looks familiar. I have seen her quite frequently. She seems to be deriving a certain pleasure from her present activity. She has a striking figure. The men are also familiar. Only this morning I had a lengthy conversation with one of them. They all appear to show a complete disregard for their safety by wrestling on the bed, despite the creaking floor. Notwithstanding Arthur's assurances, the floor could cave in at any moment. The 200-watt lightbulb illuminates the men as they one by one evoke a look of startled recognition on the woman's face . . . the look seems to say: Ahh . . . here it comes again . . . the familiar ecstasy.

The combination of the chlorine and the iodine vapor has greatly increased the sensitivity of the photographic plate. For the first time her emotions could be clearly assessed on the enlargement, but now the process took much longer. Still, I felt it was definitely worth the effort . . . all those weeks spent in the darkroom. It had taken me two years to produce the plate of Mildred. I now somewhat regretted having used Mildred instead of Muriel, because Muriel was less outgoing and more resistant to pleasure, consequently the evaluation of her troubled mind would have been more difficult, and as a result my work would have been more satisfying. After all, the conclusions I was able to draw from Mildred's photograph did not radically differ from what I had been told about her. All the same, correctly interpreting Mildred's uninhibited acceptance of F enabled me to eliminate much of the guesswork in the darkroom. The next plate, I surmised, should not take more than six months, one year at the most.

Let us now consider the world open to us, said Mildred in the adjacent room as one by one the men parted her glorious legs. In the beginning the long U-shaped corridors outside had completely disoriented them. They had lost their sense of direction as they stared at Mildred.

All this time Mr. E. Batch does not suspect a thing. The white legs that had given me so much trouble in the darkroom were now firmly locked around Arthur's obese body . . . a certain trembling motion was to be detected with the naked eye. The motion improved the shape of their bodies, I decided, since the few minor flaws became less and less evident.

What did you do this afternoon, asks Mr. Batch. Seemingly unperturbed, she crosses her legs. Mr. Batch looks at her legs. In the newspaper he has read that a certain lecture will be given at the Felt Forum, and that there is a sense of great expectation amongst the foremost scientists and amateur photographers. Mr. Batch marvels at everything Mildred says. He marvels at her excellent taste whenever she buys herself a dress. He also marvels that she had married him.

You are my developer, she says, and he beams happily.
Poor Mr. Batch.

When I entered the large hall of the Felt Forum I was greeted by a standing ovation. With quick steps I made my way to the stage, holding Mildred's photographic plate in one hand. I had, as a matter of fact, anticipated the applause, and kept my gaze fixed straight ahead as I walked to the center of the stage. It was a Tuesday, and it wasn't raining. From the amount of applause I could only assume that the hall was filled to capacity. If Arthur was to be trusted, the Felt Forum was in no immediate danger of collapsing. Somewhere among all those people waiting to hear my lecture was a face I was bound to recognize. I had taken the escalator to the second floor. It was not the first time that I had used an escalator. Naturally I was somewhat intrigued by every innovative mechanical breakthrough. A slow-moving staircase. Quite ingenious. In everything that was powered by electricity I saw a future use that could be applied to my photographic explorations. Somewhere in the days ahead I might well utilize the escalator. In the first row sat a very attractive woman. She resembled Mildred. When I took another look at her I realized it was Mildred. I knew so much about her. I could only assume that the man sitting stiffly at her side was Mr. Batch.

The entire world is now open to us, I told a hushed audience.

Mildred's husband has a neat and well-groomed mustache. The information available to him at this stage can be said to encapsulate his love for Mildred and his need to be punctual. Every question he has ever asked has been answered to his satisfaction. There were 1,855 seats in the hall. They are all occupied. He and his wife occupy two seats in the first row. He has questioned Mildred about the seats. The front-row seats seem a trifle conspicuous to him. In the past his seating experience has been a more modest one. How did you manage to get two seats in the first row, he asked Mildred.

She crosses her legs.

She also stood up frequently in order to see everyone seated behind them. Her white teeth are very much in evidence. She is smiling at the entire world.

Why on earth are you standing, asked her husband.

I'm looking for Paula.

Paula. I didn't know that Paula was coming.

Mildred wore a red suit. It was a bright red, and when she stood everyone could see her clearly, although the suit obscured certain parts of her body that could be seen on the enlargements.

As soon as electricity was discovered, the electric company was formed, I said. One of their first actions was to rush a wire to my darkroom.

Why aren't you more pleased with your success, asked a colleague afterward.

Basically I think that Mr. Batch pushes his admiration for Mildred too far. He doesn't even know that certain words he uses at night are now defunct.

Arthur shaved carefully before leaving the apartment at six. Feeling an uncontrollable yearning for F he made a pass at Muriel. She removed her false eyelashes before calmly stabbing him in the palm with a carving knife she carried in her purse. He was transfixed with surprise. Since this happened on the escalator, he disappeared from sight at the top of the second floor. Muriel in a state of shock rushes to the well-lit powder room. Luckily her purse also contained the fourth volume of the *Encyclopaedia Britannica*. She looked up fucking. It was sandwiched between "Fuchsin" and "Fucoid." It was the right book for an emergency. This could not have happened before the advent of electricity and modern photography.

I still can't get over Muriel's legs, said Arthur when he returned to the apartment, his right hand in a sling. They remind me, he said, of my recent trip to Ireland. The white legs of Ireland.

When I addressed a gathering of 1,855 scientists and photographers at the Felt Forum, the photographic plate was on the lectern, but when I returned after a ten-minute intermission it had vanished. Two years of research thrown to the wind. But two hours later I experienced my first erection in years. What a relief. I rushed to a phone, but Mildred and Mr. Batch had left for Switzerland. I think you need a new pair of pants, said Muriel when she saw me . . .

Dear Muriel. It's either that or . . .

Or what, she asks as she rushes into my arms.

WOLF'S HOUR

TENNESSEE WILLIAMS

In that old spider web which is the loom of your heart
you must somehow contrive to spin together time and peace
and somehow make it seem fitting.
      The spider's nameless,
        unknown though most familiar:
          perform in solitude, soundless, lightless,
his task: appointed by whom?

      After such abstractions as time and peace,
why not say God the Father?
        And, for apology to the skeptics
and to the ones disinclined to tolerate your attitude of romance,
      Say it's three A.M.
after an hour's sleep and a blond youth who declined to stay with you.

      Wolf's hour of night
is not well-spent alone.
        Premature contractions of a heart-valve
accent the dark outside and the shut windows on the clamor of the
        Collectors of the debris,
brute anguish of trucks clearing the street by suction.

      But in your hands' curved remembrance
the unclothed flesh of the youth who refused to stay longer,
        and you could settle for less.

      God knows if not unknowing.

FEMALE SKIN

STEVE KATZ

I made the first incision at 7:15 A.M. It was an important step. I touched the point in lightly just under the left armpit and slowly pulled the blade down along her side. The knife moved like a kayak. My system was to accomplish the separation with one continuous lateral cut, dividing her skin in two halves, front and back. That was better than a bilateral division along the axis of her symmetry because it would make for the least disfigurement when the skin had to be replaced and would permit easier disguising of the adjustments and stretching necessary to fit that small skin over my relatively large frame.

Wendy relished the attention. Her only request before I began was that I fill her navel with the six spoons of cocaine she had bought the day before in anticipation of our experiment. She interrupted me now and then as I carefully followed my procedures to instruct me where to apply the cocaine. I moistened my finger and dipped it into the heap of white powder in her belly button, rubbing it then onto her gums, or her clitoris, or touching a bit to the slight, clean, necessary wound with which I was circling her body. I was beginning to feel good.

I had anticipated some difficulty when it came to separating the skin from the body, but not so. A slow, patient tugging did it. Wendy claimed that it felt good, like a total mudpack drying.

Patience was the key to this step in this process because a slight rip of the skin would be unsightly, and obvious in public. My anticipation of tenderness at the breasts proved unwarranted also. They popped like grapes out of the peel, and Wendy sighed. I wished I could feel what she was feeling.

I laid her front skin over the dressmaker's dummy Wendy had covered with *découpage*. I had to get the back skin off in a hurry, before the front dried and lost all its natural adhesion. That's why I botched it somewhat, causing these thin patches around the buttocks and shoulder blades. I have to be very careful.

When Wendy looked at me dressed up in her goods she started to giggle.
"You look super."

She didn't look bad herself done up, so to speak, in just her musculature. Lovely spasms played across her sheets of body muscle. There was something of her revealed in this condition that I had never noticed about her before, some ephemeral disposition to grace, a lovely mild tidal mood of physicality, inarticulable, that made me want her even more than at the outset. I hadn't the slightest notion of what I was up to.

She stared at me for a long time, and then slowly began to laugh. I had never seen her find anything quite so amusing. Her laughter was natural and free, not the painful, perfunctory laughing noise she usually produced when she recognized in conversation something she understood intellectually to be funny. It was a lovely sequence of muscle spasms from her belly muscles to her face.

"I've got some rules, you know. No praying allowed in my skin. No sneezing. No spitting, eulogizing, kneeling, worshiping, howling. No sleeping permitted. No innoculating, talking, excommunicating, harmonizing, escaping, catching. Running is strictly forbidden. No smoking. No fucking. That's a poem by Nicanor Parra. Do you like it?"
"Wendy, what a time for someone else's poems."
"What would you have preferred, some bullshit emotional outburst?"

The trace of cocaine left in her navel had begun to affect me. I could see very clearly what was happening. It was confusing, but if it had been accomplished according to some scheme, that scheme was brilliant, perfect, infinite in complexity, interminable. I was restless to do something. I asked Wendy to lie as still as possible, slipped into my clothes, and stepped out the door.

It was 9:37. I was in the street, decked out inconspicuously in the absolute skin of Wendy Appel.

What's the matter? What's the trouble? What's the problem? Where's the fire? Who's the culprit? What's the reason? Was it easy? Where's the money? When is Wednesday? Who you dating? Can he have it? Do you know him? Did you touch it? Is it poison? Does it hurt? What's the meaning? What's the trouble? What's the problem? What's up Doc? Where's the contest? Are you angry? Is he happy? Did you touch it? What's the answer? What's the hurry? Was it painful? Are you touching? What's the reason? Is he singing? Were you running? Did he grow? Did she answer? Who's the culprit? What's the address? Is it money? Was he angry? Is she waiting? Is it colder? Did he take it? Does it hurt? Was it poison? Where's the honey? Are you running? Who's the boss here? Is he starving? Did you touch him? Was he up here? Is he down there? What's the matter? What's the trouble? Does it tickle? Is it empty? Does he have one? Did you touch it? Where's the fire? Are you staring? Are you nervous? Is it tender? Is it tricky? Are you happy? What's the matter? Is it over?

In the paragraphs that follow I shall attempt to answer these questions one at a time.

The problem is how to deal with all this experience. It's so bizarre. It's so banal. It's full of roaches. It's full of mushrooms. It loses so much in the telling. It's much better as a story. It's good to get it off my chest. These are things I'd rather keep to myself. Publishing certain of these facts would hurt some people's feelings. Nobody will be affected by anything I have to say here. It's not very interesting. It's absolutely hypnotizing. There isn't space or time to fit everything in. This is all I've got to say just now.

One evening after we worked all day outdoors Jingle and I both collapsed on the couch in front of the T-V. I asked if she wanted to hear part of the new story I had begun writing. She said, "sure." I got my notebook and began to read the first part of *Female Skin*. On Jingle's face, as I read, was that expression of discomfort she always gets when something seems unpleasant to her, as if all her features have been twisted about five degrees counterclockwise.

"That's the most perverse thing I've ever written," I said.

"It's gruesome," she said. "I mean I know it's not written to be gruesome. It's so matter of fact. But I always respond to the gruesome. I guess I don't like it."

She didn't ask me who Wendy Appel was. Perhaps she assumed Wendy was a fictional character. Perhaps she was being discreet. Sixteen years of a kind of marriage, and I still don't fathom her moves. If she had asked about Wendy what would I have said? "O she was just somebody I knew in Hollywood when I was working on *Grassland* out there." It's difficult for me to remember, myself, who Wendy Appel was. She was important to me when I was in that Hollywood sump. Now I no longer know her. I have to look for her in my journal from that time.

Journal From That Time

August 16, 1971. Arrived at Pierre and talked with a young blonde chubby midwestern girl on her way to Minneapolis from Colorado for a family reunion. Leo arrived with a slim, schizzy thorobred named Wendy Appel . . .

August 17. . . . we drove to Rosebud to look for Wendy Appel's acquaintance, Leonard Crow-Dog. He's a road-man for the Native American Church . . .

August 20, Hollywood. Leo is breaking his ass to make this film, popping pills, alienating his crew: these four girls—Wendy Appel, Diane Haggerty, Linda Sampson, and Dee-Anne ?, whom he expects to serve him like Playboy bunnies, are funny in their various reactions. Wendy sticks out her tongue at him (or some such similar gesture of rebellion) . . .

August 21. Last night Wendy Appel stopped by for some herb tea on the way to her date with a Japanese gentleman she'd picked up. We sat staring at each other for some long moments and then she told me she saw my face go through changes, as you do with people you have known in other incarnations. She was tired. I did for her the back rub. She explained her S & M scenes to me, telling me that what she wanted now was something from the heart. The S & M, she explained, got so repetitive. It was uninteresting to be thrown back to your childhood, to your relationship with your parents. "One day," she said, "I took the manacles and just threw them in the river." She got the hard women's lib surface over the sweetness and charm of a nice & gentle lady over the tough underbelly of an expert with a whip.

August 24. Last night I woke up out of my day of fasting through to a heavy, soulless anxiety. The notion that this Wendy Appel was "doing a number on my feelings" stuck a ridiculous thumb in my gorge. How quickly in a situation like this I build up affections and dependencies and lay out a vulnerable surface. I'm a glider. I'm a hawk. I'm the tissue of Wednesday morning. I'm the old man who wore this hat. I'm the slippery dreamy mucus of a Dunvegan frog. I'm the reptilian skin on the neck of the man who just sat down beside me at the Continental Hyatt House. I'm surrounded here by people eating Danish pastries & twisted crullers for breakfast.

Why, I woke up wondering, after such a nice day of exchanging mutual affections, and nice measures of shared feelings, after talking about preparing herself to see me at night, for fucking, for the wonderful body to body fulfillment of all our touches, long meaningless looks, secret rubs, affectionate assurances, winks, partings of lips, widening of eyes, lowering of lids, stroking of arms, why did she disappear last night into telephone silence) Where did she go? Am I making all this up? Am I uneasy & vulnerable out of my position in this movie as the secret creative flunkie? If I'm going to stick with this movie and learn what I need to out of it, I can't afford to stock so much emotional chattel. It's totally silly. Wendy Appel is a childish sweetie, who still lives on that imaginary trapeze network that occurs to us in our youth is a possible and desirable order of life, thinking that all one has to do is grab hold of the bar and let himself swing out, and between here and there—ecstasy. She lives, at least, the pattern of that autoerotic life style, though she knows better. How cruel her surface can be. Liking her is like being an aging fag fond of a mean sixteen-year-old boy. I could slip between the covers of remembrance of things past and snooze.

August 25. That emotional thrashing I laid into myself with little Wendy resolves itself this morning with forgiveness & a hug in the office, after she apologized for whatever went down, whenever that was. This is like being on a voyage, stuck on a ship with those few people with whom you form those intense seaworthy alliances. The spectacular thing is that we haven't even made love.

August 27. Ross (he came to L.A. to do some research at Great Bear observatory) suggested last night that Wendy is playing her S & M trip on my head, being so slippery, coy, evasive. It's probable that in her own head she isn't being that way at all. Certainly the trip I'm freaking off her is more like that. I'm half digging it, so it's quite possible I'm laying it quite onto myself. If nothing lubricious and loving happens this week end I'm gonna forget it.

August 28. We're gonna go to the observatory with Ross (so glad he showed up in L.A.) and look at the sun, and then we're heading for Laguna Beach, to stay with the Sukenicks, Wendy and me. Ross gave me a little instruction of the ways of a Leo lady, because Miss Appel has me delightfully confused lots of the time. It's great because it's like being with a self-propelled, self-motivated bunch of skits all the time.

August 29. . . . hear Wendy in the shower singing "Reason to Believe," Tim Hardin's song. It seems appropriate to this little bullshit experiment in emotions I'm going through with that one. She could live with a junkie. She's laying an otherwise peculiar number on my head & thass all right ma, cause I'm ridin'. I've got to tie her bathing suit now. Lord have mercy.

August 30. Back in the Los Angeles. Actually last night wasn't so bad Wendy-wise. She did try to get it on with me. She's just so twisted by whatever heaviness her old man had laid on her that she couldn't make it, but for a few minutes of desperate imitation thrusting. Then some long talk about "the problems" in which she exercised the truth, and I did the understanding, though occasionally we switched roles. Her relationship with him was a strange one. You can't build a realtionship on fantasies. Acting it out for each other continually is bound to drive you away from the source of your emotions and your mutuality. Their S & M, pleasurable as it must have been, turned to some barrier of demons to their love.

It occurred to me at one point that if I'd beat her up she might have fucked better, but since my German girl of the belt fantasy I can't get into that part of my psyche. I mean, she's so nice, I don't want to hit her.

Such swift shapes of flesh here in California.

September 1. Had dinner last night with Toni Martin, sweet black lady who works the switchboard at Park Sunset . . . Got back to the icebox and Ross, who had taken Wendy out to dinner, was sitting with her in the parking lot. I really do love her for whatever that can mean. Spent the day looking for some props with her, in the presence of a startling loveliness.

September 2. Read to Wendy from *Saw* last night till my throat was empty. She had a special smart response to the work, clearly detecting my insecurity in it, or calling the self-negating way my work has of proceeding—insecurity. She sure is right from a certain point of view.

I certainly feel touched by something in that person. She comes on me at a time when I'm vulnerable, giving myself to this unholy movie project, ready to involve myself in an immense overhaul of my work. I need her presence of mind—"marvelous, feminine & tough." I know, probably more clearly than she does, that we won't see each other much when we're all back East. Letting myself feel so strongly about her is like making myself willing to receive wounds, the compensation being what I learn about myself in her presence. I work something so similar on her. It's important for both of us in this demi-monde of costume facade & plastic to have something real to hold on.

September 3. Wendy on the plane to S.F. She's probably there by now. I left her at the airport. "I hope I see you again some time." O how she has moved me. I don't know what it is. Have I done it to myself? It's hard to know if sorting through her hang-ups and her sickness I've come to anything. Sure I have. Tough-mindedness. I love her for that. Yet I left the airport after kissing her, after all that affection, and I thought, "What was that I did to myself?" It's peculiar to think that I won't be as lonely now as I was during the days I spent with her. What a peculiar notion. Perhaps it's that we never really made love, that I couldn't ever really please her, move her, and that she never really got close to me physically, made it seem that a large part of me was with no one. Running her around for days on *Grassland* errands waiting for some small contact, some sign. And yet she was always there. Leo, her sign, the sun sign, being with her like measuring the sun. Like trying to be with sunlight. It sure is interesting. I wonder will I see her again? Seems inevitable.

These journal entries don't give me a sense of who Wendy Appel is, but I can recall what I was like when I was enveloped by her. That interlude doesn't even seem like it had duration in the past.

It's some scenario (everything in Hollywood is like some scenario) on which I tested for a particular role which didn't really suit me. The director shrugged, and I left. And I can't imagine how Jingle will react when she reads all this. Perhaps it will be as it was in Ithaca once when I drifted home late after an evening stoned with Chuck Ross and she came around the corner blindly furious and pummeled me with her small soft fists, her blows landing on my body like heavy rain. As I write she enters my studio with a cup of tea. Together we watch the house sparrows that have occupied the birdhouse on the sycamore across from my window. The male is puffed up, round, like a ball of grey cotton. It makes a racket. The female comes out. He mounts her for an instant. She goes back in. He follows her. He rushes back out and turns from that overall deep rusty brown to a grey puffy ball of noise.

"It makes him look like a different bird," I say.

"That's just probably part of their courtship ritual," she says.

The other evening I told Peter Campus and Charles Ross about this new work I've begun. Peter had described his new video pyramid piece, Ross had talked about his new drawings and his light absorption piece, so I figured I'd better get a plug in for the continuing works of Steve Katz. I told them about the incision, and Wendy's skin, and the cocaine.

"I'm glad to see you're doing something to get on top of the war," says Chuck.

"What do you mean, war?" asks Peter.

Chuck leans towards Peter and Tannis. "Maybe you happily married people aren't aware of it, but there's a war on between men and women this year."

Peter grins. "I guess we've been pretty happy this year," says Peter. He looks at me. "You know Aztec priests in certain rituals flayed young virgins and wore their skins."

I couldn't tell if that piece of information was relevant to my case.

"There's even a scene in which I go to see you," I tell Chuck. "I'm dressed in Wendy's skin."

"I hope I don't notice any difference," he says.

"The last line before I leave you is, 'He didn't notice any change in me. I didn't notice any change in him.'"

Actually one of the few times in my long friendship with Charles Ross that I was ready to forget about him was when I found out that he was "courting" Wendy Appel. That really pissed me off.

And I read the opening passage of *Female Skin* to my friend Annie Hickman the other afternoon, and she said, "That last line is really a good line."

It was 9:37 A.M. I was in the street, decked out inconspicuously in the absolute skin of Wendy Appel. I had given no thought to what I would do once I was wearing the skin. I try not to live that way. I try to think out my moves and know where I'm going. Not the case in Wendy's skin. I was lost. My first impulse was to get out of the city, go home, away from the longest avenues in the world, into the woods, where the morels might be up, and the perilous Helvella. But I couldn't go home dressed in Wendy's skin. I had to stick it out.

It was a bright clear sunny New York day, a rare one. I headed up 23rd Street. How to describe this rush of energy? Was it from Wendy's skin? I felt it tighten against my body, especially at the forehead. Something passed over the sun. A dim, distant honking sound echoed above the din of the city. I looked up. It wasn't clouds. Great chevrons of geese were crossing the sky. They were gathering like a storm and darkening the sun. No one had ever seen so many geese at once, and never in the New York City sky. In this other skin I couldn't tell what was happening. I wondered if Wendy might have understood it.

It was like a solar eclipse as the great packs of geese gathered above the city. Lights went on. Gooseshit fell everywhere, the streets so slick with it I had to skate up 23rd Street. People hid in doorways, or brought out umbrellas and tried to keep going. I kept going, protected by Wendy's skin.

Suddenly the geese all settled down around me. Huge Canada Geese, little Brant Geese and Barnacle Geese. Snow Geese. Blue

Geese. They all discovered the pavement at once, and with them, as if on their backs, they brought down the sunshine again. The pavement steamed underneath them. I continued up 23rd Street, followed by the huge contingent of geese. I moved forward, propelled by their honking. Each time I turned around they stopped and settled down, and watched me, as if I were supposed to say something. Was this the ritual of the geese? Did Wendy Appel have prior knowledge of this and not inform me? Did it happen to her every day? Or was it a unique experience that occurred only because I had put on this fascinating skin?

It was 10:37 A.M. People around me were snatching up geese, wringing their necks, tossing them over their shoulders, and hauling them home. There was nothing I could do about it. Was I responsible for the welfare of all these geese? I saw Peter Campus coming down 23rd Street. I couldn't believe that he would recognize me.

"Steve," he said, as he got closer. "You really look strange. The city must be getting to you."

I couldn't tell him about it. If he didn't notice himself what was happening it would never seem real to him.

"I'm so depressed," he said. "I've been like this for days. I'd better not talk to you. I'd better not talk to anyone."

Fuck you, depressed, I thought. I feel great. I'm moving into a whole new generation of personal experience. I can totally recollect my past, give it a good scrubbing, and store it in the closet. Peter and I parted company. He walked away slow, booting the geese out of his path. I walked away quick, listening to geese shuffle in my wake.

I led them up Seventh Avenue. It was 11:54 A.M. All traffic and commerce was halted on the avenue. As far as the eye could see my retinue of geese stretched in an endless waddle. I led them without strutting. At least a thousand geese kept themselves at all times between me and the police. Was I flying? Not really. Was I flying? Not in the sense of "flying."

At 1:17 P.M. we were at 86th Street. The mayor heard of our coming long before we arrived. He was already in flight to a different island. The palace guards gave up immediately when they saw our disciplined multitudes. We went in through the windows. We went in through the chimneys, through the drainpipes, through the doors. We occupied every room. We surrounded the grounds. The numbers that wouldn't fit I dispatched through the air to City Hall. It was 3:43 P.M. We had the city under our control.

I knew I had just committed the one most meaningful act of my life. It was my only exclusive act of power. It was a coup. My one confusion, and this was purely a personal contradiction, was that I had accomplished it in Wendy's skin. Could I have done it if I hadn't been in that condition? That question will haunt me to my grave. My geese (our geese) at any rate were now running the city. I had no reason to hang around. If you want to see them you will have to go to their offices in City Hall or visit them on the grounds of Gracie Mansion. I recommend them to you. They are honest, and mild, and willing to listen.

It was 5:30 P.M. I took the subway to Wooster Street to visit Charles Ross. At 6:15 I walked into his studio. He is working on a brand new piece, and I tried to look at it. I'm not at liberty to divulge the manner in which the piece is made, but I can describe for you some of its astounding effects.

This new work of art by Charles Ross has the capacity to cause the slow dissolution of whoever views it by irresistibly absorbing the light that makes up the viewer's own real substance. It is delightful to watch art-lovers disappear as they stand and try to look at this unique new masterpiece (I call it that advisedly).

Just before they are gone some of them suddenly understand that we are really composed of this energy we call "light." It's all we are. You can see in them that sudden instant of illumination, an impressive flash in the pan, and then they disappear. This new work of art is perfect, because it is inimitable. Anyone attempting to look at it long enough to learn how it's made inevitably disappears. This is not a warning. This is a song of praise.

We look at some photographs and drawings of the new work. I feel it grab at my eyes from the wall. I'm not afraid of it, but for the time being I think it's my obligation not to look at it.
 "It must change you to live with this new work," I say.
 "It's powerful," he says.

"Do you notice any difference in me?" I ask. "I'm wearing Wendy's skin."

Chuck shrugs and puts on his coat. "No I don't. Let's go have dinner."

He didn't notice any change in me. I didn't notice any change in him.

It was 11:17 P.M. before I got back to her place again. Wendy had been sleeping for most of the time I was gone, and so had moved very little. I woke her up. I think she smiled. Without saying a word I began to replace her skin. Then I tried to tell her all about it.

". . . incredible things were happening. I was on . . ."

"O you don't have to tell me. It's so boring. Don't you think I know what it's like to be in my own skin?"

There was silence. We stared at each other. We stared beyond each other. I knew I was in the presence of a total stranger.

"What now?" I asked. It is twelve midnight. Now it is her turn to perform her operation on me.

SEVEN POEMS

ALFRED STARR HAMILTON

DARK FLOWERS

Know of a stonewall that has windows?
Know of pulling the shades down?
Know of a stonewall for dividing our lives from the lives of
    the poor?
Know of cement leaves?
Know of the flowers that have puzzled ourselves all of our
    lives long?

ROAD TO HEAVEN

But didn't you know the road to heaven
was covered with stumbling blocks
and cobblestones
and what one cobblestone
said to the other cobblestone
and that you stopped at a cobbler's along the way
and what one cobblestone
said to the other cobblestone
and that was more important
and the blue sky was ahead of ourselves
and the road to heaven

TO HAVE AND TO HAVE NOT

with or
without gloves
some peoples have gloves
some peoples have lost their gloves
one person I know of looked behind the sofa
some peoples are like the three little kittens
who have lost their mittens
who have found their gloves again

POOR

being poor
over the bare pavements

being poor
over a bare floor

until as ever there is something
left there to be had

NOTEBOOK

I know of the bare pavements
I know of those bare stonewalls
I know of a bare room
I know of a bare floor
I know of a bare chair
I know of a notebook on a bedstand
I know of so little else
I know of November
I know of a bare windowpane
I know of a cloud in the sky
I know of barely a tea leaf
I know of barely a cup of tea this side of blue heaven

THE GLASS SHIP

I never forgot their glass buckles
I never forgot their leather satchels
I never forgot who packed their belongings
I never forgot their bag and baggage
I never forgot a polity glass strap
I never forgot the Navy waves
I never forgot the queen of the seven seas
I never forgot a glass house
I never forgot a glass ship
I never forgot a puff of smoke
I never forgot a cloud in the sky
I never forgot the blue sky
I never forgot the isinglass
I never forgot a real destroyer

WELLS

What do you know of others
What do you know of their dreams
What do you know of water on the desert
What do you know of water everywhere
What do you know of a cloud in the sky
What do you know of the three rivers that run deeper
What do you know of the three rivers that have cast their shadows
    above ground
What do you know of those wandering eyes

FOR A HUMAN TABLE, SEE "LAP."
LAPS COMMIT LAPSE.
SEE "THE LAPSE OF THE GODS."

MARVIN COHEN

Sometimes I use my lap as a human table. This is necessary when tea is served on a saucer but nothing else, with the cup fitted in the saucer, but precariously heaped to the top with tea. (Fulfilling the what-for in the design of the cup.)

The saucer is a kind of junior table for the cup. (The cup is a kind of primary table for the tea. But can a vessel be a table for what it contains? Or is that too much of a contradiction? I'm not eligible to answer this question, having never been either tea, cup, or saucer, or any combination of those three, in any order whatever. So I'm unqualified to speak on the matter, or on any other matter, except on the matter touching what I am, it being my being a human person, which is all I was ever destined to be, and all I ever managed to realize.)

Back to the pre-digression, and then to travel straight from it. The saucer, it was written, is a kind of junior table for the cup (which itself is a kind of primary table for the tea contained in it). But the saucer is unstable unless set somewhere; while the cup of tea is also unstable so long as its own table (the saucer) is. That's where my lap comes in, as a reserve stability agent (or, as said before, human table). If the tea isn't served on a real furniture table (generally of wood material, including "legs" that usually number four) then the human lap is pressed into service, and is fabricated to be improvised as a sort of artificial table. (But since when is the human body "artificial"? Certainly I wear no artificial limbs, they're my own. In fact, they're as natural as nature can be, since nature made them. *I* surely didn't make them. I—all of me—*was* made. That's what being natural consists in. So I'm a *natural* table, not an artificial one. These distinctions are not hairsplitting. They're *essential* distinctions. We'd be lost, without them.)

To go back to the substance that preceded the latest digression (determined not to be waylaid into still another tempting digression, and be thus seduced from the narrow narrative path which, taken straitly, would lead logically somewhere by virtue of itself or else virtually falter in its own purpose, end, and aim, meanly demeaning thereby its own means, by which is meant . . .).

So in the absence of a real manufactured table, I sit forward on the upholstered armchair, and rest on my lap the saucer-held, tea-filled cup. My torso tilted stiff; my upper legs absolutely horizontal, parallel to the floor and the ceiling, from pelvis to knees; my lower legs vertical, from knees to feet. Obviously (it goes without saying) my knees are bent. Bending is done by knees thousands of times before the total death of the whole person for whom the knees serve as parts. Much hinges on this function of the knees. But enough on that subject, whose obviousness is usually beneath the mentioning, just as the knees are beneath the head from which the mentioning would derive, as a physically mentalized act. Sometimes knees are above the head. It often happens to ladies who are being ravished in bed. Such indecent behavior, so as to be unobserved, is done in retired private quarters. Romantic images by the participants euphemize the sordid deed. The function of fantasy is often to exalt.

To go back behind this latest digression and recover the rather rambling theme. To recapitulate, before, in despair, I capitulate altogether.

The tea is in the cup. (Still somehow hot, just poured from that porcelain teapot where it was brewed in recently boiling water.) The cup is on the saucer. The saucer is on my lap, for want of a real table. My lap is on my upper legs and loins, which, in turn, are on the armchair. The armchair is on the rug. The rug is on the floor. Is the floor the final, all-containing table—the "table-of-contents" to begin and end the whole book? No, the floor is on something. That something is on something else. And that something else is on the earth. But where is the earth on? What table holds the orbiting earth? Magnetic space? And what's the table for that?

All will slip off the table, when the tablecloth is pulled. When the body dies, the table is turned on physical life, for the individual lodged within. Is there a tablet of laws on such a turned table? This is a tale or fable, insecurely set upon a table. All that's "on" is precarious. We use the term "based on." On what, ultimately, does all rest? The discourse on tables comes to perch upon metaphysics. Metaphysics is perched upon a set of premises.

On what, ultimately, is all laid? I'd like to lay my weary mind to rest. I'd like to allay my tedium; to lay a lady; to get laid.

The table is laid, for tea. The tea stuff is laden on it, consisting of saucer, cup, and tea, and little silver spoon. The table is my own lap. My lap may commit a lapse, if unsteadily it lets the tea stuff slip off. I drink the tea while it's hot. Tea is over. I stop my lap by standing up from the chair, ending the improvised "table." The empty cup, the saucer, the spoon, spill off by my act of standing, for the table is taken away from under them, and they don't even have a leg to stand on, so they fall on the floor, or on the rug. This was a lapse in decorum. I knead my lap, strait. Straightened, it was a lapse. The hostess bends down to retrieve the fallen service. I bend down to retrieve the hostess; to resume my service. Tea is over. We fornicate on the rug, through rips in our body where the clothes are worn through. A fitting ending to a tea. A table to uphold it all. A table to grunt, and bear the weight.

I couldn't bear the wait, from tea to that. My hostess bears my weight, weighted down with my wait. We bare the bearing instruments, unartificial. We make nature, after tea was made. For that we're made.

Our laps are pressed into service, but not for tea. We don't knead laps for tea-posture now. Our knees are freed. We weigh down heavy. We wear out the human race, as the table groans with our pounding. O Table, can you bear us? And bear too, on you successively from superficials deep down, the rug, the floor, the earth that spins, and the spin that spaces off? What's at the bottom, of this table business?

I now put a ceiling on my flawed talk of floors, and box the room in, with hinged doors. The tableau is based on what table? O Table of rectitude, uphold us!

IT DOES ME GOOD

J. LAUGHLIN

to bow my body to the ground
when the emperor passes I am

one of the gardeners at the
palace but I have never seen

his face when he walks in the
garden he is preceded by boys

who ring little bells and I
bow myself down when I hear

the bells approaching though
they say that the emperor is

very kind and not easily offended he might smile at me

if I looked up or even speak
to me but I believe that the

emperor rules by my humility
it is my humility that rules.

NOTES ON CONTRIBUTORS

"Life Uniforms" is one of a dozen stories included in *Minds Meet*, WALTER ABISH's most recent book, published this spring by New Directions. His novel *Alphabetical Africa* appeared last year, and *Duel Site*, a collection of poems, was brought out by Tibor de Nagy Editions in 1970.

E. M. BEEKMAN teaches German at the Amherst campus of the University of Massachusetts. His short story "Cornada" appeared in *ND24*, and his translation from the Flemish of Paul van Ostaijen's "Ika Loch's Brothel" was included in *ND21*.

MARVIN COHEN is a frequent contributor to the ND anthologies. *Baseball the Beautiful*, his most recent collection of short prose pieces, was brought out last year by Links Books. His *The Self-Devoted Friend* (1968) and *The Monday Rhetoric of the Love Club, and Other Parables* (1973) are available from New Directions.

"Victor" is the fourth excerpt from *Island People*, COLEMAN DOWELL's work-in-progress, to appear in these pages. His satirical novel *Mrs. October Was Here* was published by New Directions early in 1974. Before turning to writing full time, Dowell, who was born in Kentucky, worked as a composer-lyricist and playwright. His first novel, *One of the Children Is Crying*, was brought out by Random House in 1968.

MIA GARCIA CAMARILLO's work has appeared in two earlier numbers of the ND anthology: "Snailsfeet" (*ND21*) and "Brokennosejob" (*ND25*). Since last summer she has been living in San Antonio, Texas, working with the staff of *Caracol*, a Chicano newspaper set up by her husband, the Mexican-American poet Cecilio Garcia-Camarillo.

"I was born in Montclair, New Jersey," writes ALFRED STARR HAMILTON. "I am sixty years old. I couldn't afford a formal education during the '30s depression. I have been on the road. I have been a hitchhiker through forty-three states on no money at all. I am familiar with Salvation Army centers in that manner. I served (subservience) one year in the armed forces. I was A.W.O.L. I got a discharge somehow. I am immune."

Biographical information on JOSÉ HIERRO will be found in the note preceding his "Three Poems." LOUIS M. BOURNE has contributed his translations of the work of modern Spanish poets to several recent ND anthologies: Carlos Bousoño (*ND29*), Claudio Rodríguez (*ND26*), and Angel González (*ND28*). A former editor of *The Carolina Review*, he now lives and works in Madrid.

STEVE KATZ is the author of three novels—*The Exaggerations of Peter Prince* (Holt, 1968), *Creamy and Delicious* (Random House, 1970), and *Saw* (Knopf, 1972)—and a book of poems, *Cheyenne River Wild Track* (Ithaca House, 1973). He is currently working on some screenplays, the most recent one about Toussaint L'Ouverture.

ANDRÉ LEFEVERE teaches in the Department of German Philology at the University of Antwerp, Belgium. His biographical notes on the various contributors to his mini-anthology of contemporary German "Irritational Verse" will be found on page 102.

A group of ROBERT MORGAN's poems was included in *ND26*. He was a recipient in 1974 of a fellowship from the National Endowment for the Arts. He is the author of two published collections of verse—*Zirconia Poems* (Lillabulero, 1969) and *Red Owl* (Norton, 1972)—and *Land Diving*, his latest compilation.

The late CHARLES OLSON's diaries, *Among the Ruins*, is being brought out this spring by Grossman Publishers. CATHERINE SEELYE, who edited the volume, obtained the text from the Olson Archive in the Wilbur Cross Library at The University of Connecticut.

JAMES PURDY's stories, poems, and short plays appear regularly in these pages. His most recent novel is *The House of the Solitary Maggot* (Doubleday, 1974), the second part of his ongoing work, *Sleepers in a Moon-Crowned Valley*.

Last fall Black Sparrow Press brought out GILBERT SORRENTINO's *The Masque of Fungo: Flawless Play Restored*. Like "Art Futures Interview of the Month: Barnett Tete," it is a self-contained excerpt from his novel-in-progress, *Synthetic Ink*. Sorrentino's two most recent novels, *Imaginative Qualities of Actual Things* and

NOTES ON CONTRIBUTORS 183

Steelwork, were published by Pantheon. *Splendide-Hôtel*, his booklength meditation on Rimbaud, is available from New Directions as a paperbook (NDP364) and in a clothbound, signed limited edition.

TENNESSEE WILLIAMS's revised *Cat on a Hot Tin Roof* premièred at the Shakespeare festival in Stratford, Connecticut, last summer, and reopened in New York shortly afterward to wide critical and audience acclaim. The text is being issued this spring as New Directions Paperbook 398. *Eight Mortal Ladies Possessed*, his most recent collection of short fiction, was brought out in September 1974.

Biographical information on STANISLAW IGNACY WITKIEWICZ ("WITKACY") will be found in the note preceding his play "The New Deliverance." DANIEL GEROULD teaches theater at the Graduate Center of the City University of New York. He has published two volumes of translations of Witkacy's dramas, *The Madman and the Nun* (University of Washington Press, 1968) and *Tropical Madness* (Winter House, 1972), and is now completing a study of the Polish author for Twayne. JADWIGA KOSICKA, his co-translator, was born in Poland during the Nazi occupation and now makes her home in New York. With Gerould she is preparing a new collection of Witkacy's plays, *The Beelzebub Sonata*, as well as a translation of his treatise on drugs.

"Entropisms," a series of prose poems by HARRIET ZINNES, appeared in *ND27*, and further selections have been published in *The Carleton Miscellany, Choice*, and *Confrontation*. An associate professor of English at Queens College of the City University of New York, and the author of two volumes of poetry (*Waiting and Other Poems* and *An Eye for an I*), she is also a widely published literary and art critic.

NEW BOOKS
BY RECENT CONTRIBUTORS

Spring 1975

MINDS MEET / Walter Abish. Can fiction explore itself? In Walter Abish's first collection of short fiction, each of the dozen absurdly humorous yet frequently disquieting stories provides the material that probes its own existence. Over and over again, the author playfully undermines familiar meanings and communicative signals with the unexpected, often bizarre juxtaposition of image and word. "Rimbaud talked about the 'Alchemy of the Word,'" writes critic Anna Balakian. "Abish's use of words does not transform reality but anchors it with a funny and frightening firmness." Available clothbound and as New Directions Paperbook 387.

THE FALLING SICKNESS / Russell Edson. The first collection of short plays by the author of *The Very Thing That Happens* (NDP137) will be welcomed by the extensive and devoted readership he has gathered over the years. His turn to the theater reveals a new dimension in his work, a vein of dramatic satire that makes all the more vivid the nightmarish absurdities which underscore so much in our lives. Available clothbound and as NDP389.

CAT ON A HOT TIN ROOF / Tennessee Williams. One of the playwright's most popular successes, *Cat on a Hot Tin Roof* was rewritten and produced in the summer of 1974 by the American Shakespeare Theatre in Stratford, Connecticut, with a new third act, along with other substantial revisions. Receiving high praise in all quarters, it reopened on Broadway in the fall. The text of this, the new version of the play is being published for the first time. Available clothbound and as NDP398.